Census 2020

Teresa A. Sullivan

Census 2020

Understanding the Issues

 Springer

Teresa A. Sullivan
Department of Sociology
University of Virginia
Charlottesville, VA, USA

Interim Provost and Executive
Vice President for Academic Affairs
Office of the Provost
Michigan State University
East Lansing, MI, USA

ISBN 978-3-030-40577-9   ISBN 978-3-030-40578-6   (eBook)
https://doi.org/10.1007/978-3-030-40578-6

This Springer imprint is published by the registered company Springer Nature Switzerland AG
The registered company address is: Gewerbestrasse 11, 6330 Cham, Switzerland

Preface

As a graduate student at the University of Chicago, I had the privilege of working with the late Professor Philip M. Hauser. Before his career at the University, he had been a professional at the U.S. Census Bureau, and he was briefly the Acting Director for the 1950 Census. In the course of my doctoral research, I had the good fortune to accompany him on a research trip to Asia, where he spoke with the officials in seven national statistical offices. I was able to see firsthand how the census process was part of nation-building and how the statistical infrastructure of a society helped it economically and socially.

My first professional position was at the University of Texas at Austin, where I was affiliated for many years with its Population Research Center, which had an extensive library of international censuses. I had the opportunity to observe scholars who used census data in their research and cared about how census procedures could be improved. There were many good demographers at Texas, and some of them who had a lasting influence on me were Professor Harley L. Browning, Professor Dudley L. Poston, Jr., Professor Omer R. Galle, Professor Frank D. Bean, and the late Professor W. Parker Frisbie.

During my career I had the opportunity to provide service to the Census Bureau. From 1989 to 1995 I was a member of the Census Advisory Committee on Population Statistics, which I chaired in 1991–1992, and from 1992 to 1994 I served on the National Research Council's Panel on the Requirements for the Year 2000 Census and Beyond. Through these opportunities I was able to learn much more about the workings of the census, and I met many scholars whose work influenced mine. I would mention particularly the influence of Professor Reynolds Farley of the University of Michigan and my graduate school classmate the late Professor Clifford C. Clogg, then of the Pennsylvania State University. Dr. Connie Citro at the National Research Council was an insightful guide to the broader statistical community in which the Census Bureau operated.

Beginning in 1995 I became a full-time administrator, holding positions at the University of Texas-Austin and the University of Texas System, the University of Michigan, the University of Virginia, and Michigan State University. At every stop I worked with many wonderful, dedicated people who committed themselves to the preparation of the next generation. My background as a demographer was always helpful to me, and I gained an increased appreciation for the difficulties of steering large organizations in making the right choices. Universities, like the Census Bureau, are always balancing consequential decisions and seeking data on the possible effects of those decisions.

This book grew from my association with Professor Karen Kafadar, Chair of the Statistics Department at the University of Virginia and immediate past chair of the American Statistical Association, and with Laura Aileen Briskman, editor at Springer, who believed that the census story was worth telling. My husband, Professor H. Douglas Laycock, a constitutional scholar, led me through many fruitful discussions of how the constitutional mandate affects our politics as well as how the census affects our economy and society.

I cannot say enough positive things about my research assistant, Sarah Panuska, and her diligent work. Catherine Burns offered me valuable editorial support.

Finally, I am indebted to the many Census Bureau professionals with whom I talked and corresponded over the years. We did not always agree, and the view from the academy looks different from the view in Suitland, Maryland. Nevertheless, I have the greatest respect for their professionalism and dedication, which is usually unsung and even unnoticed. Whether or not they would say it, they are important contributors to the project of American democracy. They will not like everything I wrote here, but I hope they understand that I have great respect for them and their work, and I affectionately dedicate this volume to them.

East Lansing, MI   Teresa A. Sullivan

Introduction

It will take the average American household ten minutes or less to complete the 2020 Census questionnaire. That simple action will set in motion events that will affect that household for a decade, from the Congressional district in which the adults can vote to whether the local highway gets repaved. Census data will be important planning tools for the children's school district, for emergency responders, and for hospitals. New businesses will make decisions about their preferred locations based on the data, and marketers will decide what coupon offers should go to the mailing addresses in that household's ZIP code. Whether a new Asian market or a McDonald's builds nearby, whether the school district hires extra bilingual teachers, whether the city council decides that the tax base is growing and so the tax rate need not increase, all these will be influenced by the information provided from millions of census returns.

Completing the census questionnaire is required by law—it is that important. But more than that, information from the census is part of an accessible but usually unseen infrastructure of data. This data infrastructure undergirds our government and our economy.

This data infrastructure also undergirds our democracy. A government that is "of the people, by the people, for the people" needs to know who the people are, where and how they live, and what they need. One of the great compromises in the development of the U.S. Constitution was that each state would get two Senators, but the lower house of Congress would have representatives elected in proportion to the number of people enumerated in the census. Today there are also many federal funding formulas that depend at least in part on data derived from the census, with as much as $1.5 trillion at stake.

So, the ten minutes that the household takes to complete its census questionnaire is consequential. And because it is consequential, people care about the census and there are controversies associated with the census. This book examines some of those controversies and their consequences.

Chapter 1 describes the role of the census in our democracy. Chapter 2 provides detailed information about how the census is actually conducted. If you know very little about the census, you might start with Chap. 2 to get a basic grounding.

There is a glossary in the back of the volume to help you with unfamiliar terms. Chapter 3 examines the undercount and the underreporting of data, and how the Census Bureau seeks to estimate the amount of error and in some cases to correct the error. Chapter 4 discusses the two Supreme Court decisions in 2019 concerning whether the 2020 Census must include a question on citizenship (the answer was no) and whether the courts could review charges of gerrymandering (the answer was a qualified no). Chapter 5 discusses whether the information submitted to the census is safe—safe from the nosy neighbors, safe from international spies, and safe for decision-making. Chapter 6 discusses whether in this information era the census is still really something that we need, especially given its cost.

Each chapter is followed by its own references so that as far as possible the chapters stand alone. An index is provided to help look up important concepts, such as the undercount, that will occur in many chapters.

Contents

Chapter 1
The Census: A Pillar of American Democracy and American Society

Abstract Two great civic mobilizations of the American people occur in 2020: the census on April 1 and the national elections on November 3. The U.S. Census is mandated by the Constitution to occur every 10 years, in years ending in zero, to provide the numbers needed to reapportion the House of Representatives. This is also a reapportionment of the Electoral College. In addition, the census numbers are used by state governments to redraw legislative districts, and the federal government uses the numbers in various funding formulas to distribute some $1.504 trillion in funding for highways, hospitals, schools, and many other purposes. Planning for the census has been underway for many years to adopt new technology and improve techniques at the same time that the population remains mobile and much more diverse, and therefore harder to count. For the first time in 2020, online responses will be the preferred response mode for the legally required census form.

Keywords Census · Reapportionment · Redistricting · Coverage error · Response error · Online response

> The obligation to conduct a decennial enumeration of the population appears in the sixth sentence of the Constitution, as the very first duty given to the new federal government before the enumeration of legislative power, before the power to declare and wage war, before the resolution of federal judicial cases. (Levitt, 2019:59)

In years ending in zero, the United States undertakes its decennial national portrait. With its large geography of highly mobile people, great diversity in demographic characteristics, and robust political debates, this mandated enumeration is neither easy nor cheap. Taking our census is our biggest peacetime mobilization, involving hiring half a million temporary workers and seeking to count every one of us at our usual place of residence in the fifty states, the District of Columbia, Puerto Rico, and the island areas of American Samoa, Guam, The Commonwealth of the Northern Mariana Islands, and the U.S. Virgin Islands. And while the census is important for the demographic, geographic, and economic information it provides, it is most important for its role in American democracy. Put simply, everyone counts.

1.1 The Census and the Presidential Election: The Great Conjunction of Democracy

For any democracy to survive, successive generations of the population must embrace its institutions and practices. Of necessity, a democracy involves a certain periodicity, recurring events that ensure its continuity. One obvious such event is elections, of which the presidential election is the most significant. Held every 4 years, the presidential election is enshrined in the United States Constitution in Article II, section 1, as amended by the twelfth amendment.

Also enshrined in the Constitution is the actual enumeration of the population.[1] The Constitution was ratified in 1789, and the first census occurred in the next year, 1790.[2] A census has occurred every 10 years since then, with increasing sophistication and attention to error. Data from the U.S. Census serve a variety of functions, but the two most important are the reapportionment of the House of Representatives among the states and the use of census data for the allocation of over $1.504 trillion in federal expenditures through various funding formulas.[3] Both elections and the census are important to democracy because they speak to the equality of each individual with respect to the government. Both representation in the House of Representatives and receipt of federal funding are based on population counts, and every person counts.

---

[1] The census is mandated in Article 1 Section 2 of the U.S. Constitution. Since 1980, the U.S. Census is governed by laws in Title 13 of the U.S. Code.

[2] The first census calculated the population of the U.S. in 1790 to be 3,893,635. This included 694,280 enslaved people who did not have the rights of citizenship. See, 1790 Census Population Results. (n.d.). in chapter references.

[3] Reamer, A. (2019) "Counting for Dollars 2020: The Role of the Decennial Census in the Geographic Distribution of Federal Funds."

Every 20 years, the presidential election and the decennial census coincide within a single calendar year, and 2020 will be such a year. We might think of the coincidence of these two great national mobilizations as the Great Conjunction. Previous years in which the census coincided with a presidential election have been notable for events that marked great shifts in the nation. In 1820, during the Era of Good Feelings, James Monroe was re-elected without significant opponents, the last time that a president would be unopposed for election. By contrast the election of Abraham Lincoln 40 years later in 1860 set the stage for the Civil War that divided the country over slavery, led to some 750,000 deaths, and resulted in bitter economic and social tensions that continue to reverberate.

The arrival of a new century and its promise of progress came in 1900, when William McKinley was re-elected, the first president to be re-elected since Ulysses Grant in 1872. Perhaps more significant was that his new running mate, Theodore Roosevelt, would become president the following year because of McKinley's assassination. Theodore Roosevelt's fifth cousin Franklin Delano Roosevelt was re-elected in 1940 to a record-setting third term while the effects of the Great Depression lingered and World War II loomed. The Reagan era began in 1980, followed shortly by the fall of the former Soviet Union and a significant political realignment in the United States, accompanied by resurgent conservatism.

What political history of the American democracy will be written in 2020 remains to be seen. But American elections depend, in important but sometimes subtle ways, on the information that comes from the census. The census date of April 1, 2020, falls squarely during the political primary season and seven months before Election Day on November 3, 2020. And while the popular vote is important in terms of candidate credibility, the vote within each state determines which party's electors will cast votes in the Electoral College, which bears the ultimate responsibility for electing a president. Census results from the previous 2010 Census play a key role in the Electoral College because the census was used to reapportion the House of Representatives. The Electoral College votes for each state are the sum of that state's senators (always two) and representatives (at least one). Thus, the reapportionment of the House of Representatives is also a reapportionment of the Electoral College. As the census documents population growth and shifts, the Electoral College votes cast by different states also increase or decrease.

Besides the balloting for President and Vice President, in 2020 Americans will also vote for every member of the U.S. House of Representatives. With few exceptions, these representatives will be elected from districts that were drawn based on data from the 2010 Census. The 2020 Census will provide critical data for Congressional elections for the remainder of the decade. The 2020 Census will be used to reapportion the 435 House members among the fifty states, with some states likely to gain representatives while other states lose. Then, in most states the 2020 Census data will be used to redistrict – that is, to draw the Congressional districts within the state. Most states will also use the 2020 Census results to create voting districts for state legislators and sometimes for county and city council districts as well.

Voting is not required by law, and many people who are eligible to vote do not register to vote or, if they are registered, they do not cast a ballot. *The census is required by law,* and everyone must be counted. People who do not or cannot vote are nevertheless represented and so they are counted: children, immigrants who have not yet become naturalized, and some adults who have been convicted of felonies.[4]

Because of its universal coverage, the census affects all Americans whether native-born or foreign-born and without respect to their race, creed, language, nationality, gender, or gender expression. The census affects schoolchildren, workers, and the retired, and people from all walks of life. Regardless of physical ability or illness, criminal background, family background, or education, everyone counts. The objective of the 2020 Census will be to count every person in the United States "once and only once and in the right place (Groves 2010a)."[5]

1.2   The Census, Controversy, and Integrity

Taking the 2020 Census will be the largest, most expensive statistical undertaking that the United States has ever done. Because this work will be done at taxpayer expense, there will be controversies large and small over every aspect of the process. These controversies too are part of the democratic process. It is a privilege of Americans to have the ability to criticize the government freely, to propose alternative procedures for the census, and to enjoy the use of free, valid, publicly available data. Among scholars of the census—statisticians, demographers, geographers, economists, historians and others—the process of scholarly review, argument, criticism, and revision has improved census procedures and analysis.

Beginning early in the twentieth century and continuing through at least the past seven censuses, the U.S. Bureau of the Census (hereinafter referred to as the Census Bureau) has steadily improved its professional and technical prowess in all the aspects of census-taking. So well-recognized is its expertise that the Census Bureau offers advice and consulting to national statistical offices all over the world.

---

[4] The disenfranchisement of felons varies by state and by crime. All states (and the District of Columbia) prohibit voting while incarcerated for a felony offense with the exceptions of Maine and Vermont. Thirty-five states do not allow persons on parole to vote, and thirty-one of these states do not allow persons on probation to vote. See, Felony Disenfranchisement Laws in the United States (2014) in chapter references.

[5] This is a phrase that is often linked to the role of the census as it demarcated in the Constitution, although this is incorrect. The phrase has gained popularity within the Census Bureau since 2010. Former census director Robert Groves often utilized this phrase in regard to the mission of the 2010 census, including his director's blogs and other electronic letters on April 14, May 4, May 24, June 9, July 12, September 9. For more see, Groves (2010b), Groves (2010c), Groves (2010d), Groves (2010e), and Groves (2010f) in references.

But the important political purposes of the census can collide with the professional integrity of the professionals who conduct the count and analyze the results. Politicians may vary in their enthusiasm for the Census Bureau's central purpose of counting everyone once, only once, and in the right place. On its website, the Census Bureau describes itself as a non-partisan agency of the government (U.S. Census 2019). Given the close political divisions in many states and the country as a whole, however, advocates from many different perspectives have tried to have their partisan purposes bleed over into the census, seeking to manipulate the results in a way that they believe will favor their partisan interests.

As we shall see in later chapters, such issues have arisen with the 2020 census. Constitutional checks and balances are weak with respect to the census. The Constitution charges Congress with undertaking the census, and the early censuses were closely overseen by Congress. Over time, however, Congress has delegated the conduct of the census to the executive branch, specifically to the Department of Commerce.

To be sure, there is Congressional oversight of the census even though the census is within the executive branch. Title 13 of the U.S. Code governs the Census Bureau.[6] The U.S. Census Bureau is an agency within the Department of Commerce. Both the Secretary of Commerce and the Director of the Census are political appointees, nominated by the President and confirmed by the Senate. The House must initiate the appropriation to conduct the census, the Senate must approve the appropriation, and the President must sign the appropriation into law. Congress requires regular reports about the census, including a report on the questions to be asked, a report that was submitted on March 29, 2018. Nevertheless, as I have argued elsewhere, the division of practical authority between Congress and the President is not a sufficient guarantee that the best professional judgment in the Census Bureau will prevail. A single, disciplined political party in control of the Presidency and the Congress would be able to torque the census in partisan ways (Sullivan 2020).

The census is not only our decennial national portrait, it is also a treasure trove of information about who makes up the American population. This volume is intended to explain the significant issues of the census to all who are interested, whether they be from the older generation that now remembers many censuses, or the recent high school graduates who do not really recall being enumerated in 2010 and who have not yet voted in a presidential election.

While this chapter explains why the census is taken, in Chap. 2 I will explain in non-technical language the four other W's of the 2020 Census: who, what, when, and where the census is conducted. Subsequent chapters will examine controversies and issues that have arisen regarding the 2020 census.

---

[6] This has been the case since 1954, where Chapter 1, Section 2 states that "The [Census] Bureau is continued as an agency within, and under the jurisdiction of, the Department of Commerce." This was a change from 1952, when the Census Bureau was temporarily an agency within the Department of the Interior. See U.S. Code Title 13 in chapter references (n.d.).

1.3   What Is a Census?

Although censuses have been conducted in many times and places through human history, their purposes were typically to estimate military strength and to assess the extent of the tax base. Neither of these purposes appealed to the public; indeed, censuses were viewed as instruments for national bragging rights at best and for oppressing people at worst. The United States is given credit for the first modern census, taken in 1790. This census was also different in its principal purpose, which was to ensure democratic representation by basing the House of Representatives upon the population. Since that first census, there have been many improvements in census taking, and today most countries in the world take a census.

1.4   Criteria for a Modern Census

International standards for a census have evolved over time. The United Nations has identified four elements that define a modern census (United Nations Department of Economic and Social Affairs, Statistics Division 2008a, p. 5–6). The census is *individual*: this means that each person is counted and information is collected on each individual. In early U.S. censuses, detailed information was collected on the head of the household, but the other members of the household might be summarized by age and sex without being named individually. It took several decades of experimentation before each individual was named with individual data recorded. Since 1940 the United States has combined a census of housing with the census of population, and the census form is distributed to households, and within those households every individual is to be included. Individuals living alone are considered to be their own household. Individuals living in group quarters (such as prisons or college dormitories) are counted there even though their living arrangement is not exactly a household.

The census should also be *universal,* meaning that everyone within the defined territory should be counted. The defined territory for the United States has steadily expanded with the westward expansion of the country and with the results of several wars. The requirement for universality means that even the most remote parts of the country must be included. For the 2020 census, the enumeration of some of the most remote villages in Alaska and remote parts of northern Maine will require special accommodations to the long winter. The 2020 Census will enumerate the fifty states, the District of Columbia, Puerto Rico, and the U.S. Island areas (American Samoa, The Commonwealth of the Northern Mariana Islands, Guam, and the U.S. Virgin Islands), as well as some Americans who are temporarily abroad or at sea.

The census should be *simultaneous*, meaning that the data should refer to a specific date. In a large country with remote areas, it is not possible to complete the count of every household on a single date, but it is possible to specify that the information provided should be current as of the census date. In the United States that date is customarily April 1, a date that was selected to avoid the worst of winter weather and the dislocation of summer vacations. That date does unfortunately coincide with April Fool's Day, leading to many jokes about the census.

The fourth criterion the United Nations specifies is that the census be *periodic*, meaning that it recurs on a regular basis. In the United States that period is a decade, with the census taken in years ending in zero. That period is determined by the Constitution (Article 1, section 2). Many countries in the British Commonwealth have traditionally taken their censuses in years ending in 1, such as 2001, 2011, and so on. For purposes of international comparability, the United Nations now recommends that all countries take censuses in years ending in zero.

Some sources add additional best practices. One is that the census should produce *small area data*. The census should be able to produce data for small geographic areas, subject to safeguards for confidentiality and privacy (United Nations Department of Economic and Social Affairs, Statistics Division 2008b, p. 48, 76). This criterion greatly increases the value of the census for local governments, school districts, and businesses.

The United Nations also lists as a best practice that the census be published – that is, that the data collected by the government be shared with the population. In the United States, the census has been communicated through numerous summaries, reports, and books, but the preferred method for dissemination of census results today is through the Census Bureau's website, www.census.gov and robust public releases of both data and analyses.

1.5   Hallmarks of Quality in the Census

How useful a census will be is a function of its quality. Two hallmarks of quality are *completeness of coverage* and *accuracy* of the information. A complete census counts all the people who should be included and counts them only once. An accurate census has correct information upon which users may rely. These are ideals for the census, and they are never totally realized, but census personnel work very hard to collect accurate, complete data, and then they engage in a variety of skilled technical efforts to identify coverage errors and estimate the effects of errors. In their process of error detection, employees of the Census Bureau have strongly influenced the growth of the disciplines of statistics and demography and have contributed many of the methodological innovations that are today used in economics, marketing, and many other fields.

1.5.1   Completeness of Coverage

In counting a large, diverse, and mobile population of more than 330 million people, completeness is a particular challenge. Enumeration was once done only by trained interviewers who went door to door. The Census Bureau still employs trained census takers, but for reasons of cost and respondent convenience the in-person visit is not the first contact and most households will never encounter a census taker. In 2020 most households will receive their census form through the mail, with several options for responding, including telephone, a mail-back return and—for the first time—an online option. Most people will return their census forms in these ways, but the return rate has dropped from 87% in 1970 (Stackhouse and Brady 2003) to 79.3%% in 2010 (Levitt 2019). Every percentage point drop in the rate of return increases the expense of conducting the census.

As the immigration stream to the United States has become more diverse in terms of countries of origin, communicating with non-English speakers or writers is an important technique in seeking a complete count. For people who do not read English or are not comfortable with an English-language form, online and phone responses can be completed in twelve languages besides English: Spanish, Chinese, Vietnamese, Korean, Russian, Arabic, Tagalog, Polish, French, Haitian Creole, Portuguese, and Japanese (Fontenot Jr 2018). Every household receives information on how to complete the form in a language other than English.

For the people who do not return their forms, the Census Bureau will rely on follow-up mailings, telephone calls, and personal visits from census takers to complete the count. Some people move often, live in a different city from the city in which they work, maintain two homes, or are taking a long trip. Other people want to avoid contact with the government, perhaps because they are engaged in illegal activity, live in illegal housing, or are present in the country illegally. Some people just don't like the government. Census Bureau personnel will work tirelessly to discover whether a house is vacant or whether its occupants just did not or could not complete the form.

A particular emphasis of the Census Bureau is counting the hard-to-count population such as migrant workers, people in hospitals, travelers in hotels on the census date, and the homeless. Chapters 2 and 3 contain more detail on how this part of the enumeration is done.

If the census is incomplete because of individuals or households who are missed, there is an undercount. If some people or households are counted more than once, there could be an overcount. It is possible for there to be an undercount of some geographic areas or types of individuals, while there is an overcount elsewhere. Undercounts and overcounts are errors of coverage.

An undercount can result in significant loss of revenue to a locality because of the federal funding programs that are tied to population counts in funding formulas. And a statewide undercount can result in Congressional (and Electoral College) representation going instead to a different state. For this reason, state and local officials have been active in helping the Census Bureau make a complete count in their locality. Complete Census Committees have been formed in most states and many cities, and they serve a variety of purposes. They can alert the Census Bureau to new housing that might be occupied but was not in the census canvass; they can help mobilize churches and community groups to encourage participation; and they can publicize the census so that everyone understands the significance of April 1, census day.

1.5.2   Accuracy of Information

Accuracy of information refers to how correct the information on the census form is. Errors can creep into the data in numerous ways. One source of error that is within the control of the Census Bureau is *processing errors*. Processing errors refer to any inaccuracies that come from the ways in which the censuses are collected, stored, coded, or analyzed. A processing error could result, for example, from mistakes in the mail-out process, mistakes in the receipt of the completed forms, and mistakes in putting the information into machine-readable formats. Then there can be errors in processing the machine-readable data, in making tables of data, and so on. Census Bureau professionals have been very attentive to potential sources of data error and methods of checking for and correcting errors. Because of the size of the census, many innovations in data processing and computing were pioneered at the Census Bureau (Anderson 2015).

A more difficult source of error is *respondent error*, also called errors of reporting. These are the mistakes that people make, intentionally or not, in completing their forms. Sometimes these are careless errors, as when a respondent accidentally skips over a question or indicates the wrong response. In a non-family household, such as a group of roommates, the person completing the form might not know the information for other household members.

Some errors of reporting are so well known that demographers have developed techniques for detecting them and correcting published age distributions, if necessary (Shryock and Siegel 1976). One such error is *age-heaping*, or the tendency of people to round off their ages to numbers ending in 5 or 0 rather than providing exact ages. There is some association of age-heaping with more elderly respondents; in some parts of the world, age-heaping is associated with low levels of formal education. One reason for including both age and date of birth on the census is to provide a check on age-heaping.

The census conducts elaborate pre-tests and dress rehearsal censuses to try out their procedures to eliminate potential errors. Even with this testing, some people misunderstand the census questions or read ambiguities into them. As immigration from Mexico, Central America and South America became a more important part of the population, decennial censuses began to ask questions about Hispanic ancestry. One classic misunderstanding resulted in a great increase in the reported Hispanic population in unexpected areas. It turned out that a number of people in Midwestern and Southern states, not understanding the Hispanic ancestry question, reported that they were of "Central or South American origin." The pre-printed response was apparently interpreted as a query about one's geographic location (Snipp 2003, p. 575).

1.5.3   Analysis of Coverage and Accuracy

An important part of the work of the Census Bureau takes place after the census, where there is a process of continuous improvement to make the next census more complete and more accurate. Statisticians and other professionals review the data and procedures of the census, seeking to detect sources of error.

Preventing response error is difficult. Detecting response error, or at least the possibility of response error, requires the considerable ingenuity and expertise of the professionals at the Census Bureau. An important tool for detecting error and estimating its effects has been the post-enumeration survey. The post-enumeration survey (PES) is a sample survey taken shortly after the census. The respondents for the survey are not pre-selected; rather, they are selected through a multi-stage sampling process that approximates an equal chance for each household to be included. The PES repeats questions from the census enumeration and compares the answers given by the household.

The first error of significance is respondents—households or individuals—who are counted in the PES but were missed in the census. This information is important for estimating net undercount and also for understanding the characteristics of those who were missed. Some typical examples of individuals who are missed are babies and young children or members of a household who were traveling. Some typical examples of households that are missed are short-term renters, households that were moving on the census date, or households in non-conventional housing. There has been a persistent differential undercount of minorities and low-income households.

For households that complete information both for the PES and for the census, the additional information offers an opportunity to compare the answers. When an answer given in the census differs from an answer in the survey, there is a suspicion of a response error. When there is a pattern of apparent response errors then the Census Bureau devotes more effort to understanding the error and estimating its possible effect.

The post-censal analysis also involves administrative records. The census is the cornerstone of the American social statistics, but it is conducted within the context of many other types of data that are available. This system of social statistics provides checks on the quality of the census data. For example, the Census Bureau can estimate the expected number of children aged 10 and under to be enumerated by combining three sources of data: birth certificates of children who were born after April 1, 2010; data on children admitted as immigrants during the decade; and an adjustment for the number of death certificates of children. Good counts on the number of children are essential to local school districts, among other entities.

1.6 Other Uses of the Census

Besides the governmental reasons already discussed, the census serves a variety of other functions, and this fact reinforces the need to emphasize the quality of the data. Mapping every individual geographically makes census data the ideal source to use for survey sampling, whether by the government or by private firms such as political polling and marketing studies. The Census Bureau itself undertakes a number of important national surveys, such as the American Community Survey and (together with the Bureau of Labor Statistics) the Current Population Survey.

Combined with data on births and deaths from vital statistics offices, the census data are used to make population projections and to do estimates of population size between censuses. Even before the census is conducted, demographers can already estimate that approximately 332,527,548 million people will be counted, based on the 2010 count and information on fertility, mortality, and net migration in the succeeding 10 years (University of Virginia Weldon Cooper Center 2018). Intercensal estimates are provided for states, counties, and cities; the most uncertain factor in these estimates is migration, because many families move every year. The migration data are compiled from such sources as driver licenses, utility hookups, and payroll information. It seems likely that near the end of a decade this information becomes less reliable and needs refreshing from the census data.

Facsimiles of the actual census forms are currently released 72 years after the census is conducted, so that census data have become an important source of information for people interested in genealogy and history. Both the National Archives and commercial sites such as Ancestry.com make this information broadly available to the public.

1.7 The Census Mirrors Its Times

As times and circumstances change, the Census Bureau modifies the questions it asks and also changes the possible responses. For ease in completing the census questionnaire and efficiency in tabulating the results, a number of the questions asked on the census form have pre-coded answers that are easily machine-readable. There is often controversy over whether the pre-coded answers are accurate or sufficient.

One example of the adjustment to changing times was eliminating the term "head of the household." This term had by default been assigned to husbands in married couples. The term used now is reference person, and that is the person who is listed first on the form. The form suggests that this be the person who owns or rents the housing unit. To understand household composition, the census return asks for the relationship of each member of the household to the reference person.

Something that was once considered a response error can come to be an accepted choice in the precoded options. Such a case was same-sex couples, who could not be reported as married in earlier censuses. Some same-sex couples wrote in that they were married, and while that was legal in some states, their statuses were not recorded as such in the census. From one perspective, their responses could be seen an error of reporting. In 2015, however, the Supreme Court's decision in *Obergefell v. Hodges* legalized same-sex marriage throughout the United States, and so same-sex married couples will be able to report themselves as married in 2020.

In 2020 the options for reporting sex are male and female. It is foreseeable that for the 2030 Census there will be a campaign to provide additional options such as gender fluid, transgender male, and transgender female.

The census has asked about race and ethnicity for many years, and the questions and the possible responses have changed with every census (Snipp 2003). A separate question on Hispanic or Latino heritage was added in 1970 (U.S. Census Bureau n.d.-a; n.d.-b; n.d.-c).[7] But the Census Bureau has been rightfully reluctant to set out definitions of who may be counted in any particular racial or ethnic group. There is no official definition, for example, of who qualifies to be an American Indian or an African American.

Acquiescing to requests from some citizens and interest groups, respondents may report belonging to multiple racial or ethnic groups. The Census Bureau does not determine whether someone is "correct" in their response to such questions. As a result, the response error for the race and ethnicity questions is difficult or impossible to determine. In 1960, the Census Bureau nearly removed the question on race from the census questionnaire, in response to arguments that race was no longer salient. Instead, the Civil Rights movement underscored the salience of race and indeed racial differences have formed an important theme in census analyses of undercount, housing segregation, and age differences.

1.8   Counting the Enslaved, 1790–1860

Perhaps the most important reason for continuing to ask the population about race is to be able to trace the effects of slavery in the current population. Slavery was one of the great divisive issues for the United States from its founding. The issue was taken up in various ways at the Constitutional Convention, and the Constitution itself banned the importation of any additional slaves after 1808. Perhaps the most controversial part of the Constitution dealing with slavery, however, was that people in slavery were counted for eight censuses as three-fifths of a person.[8]

The Constitution in its final form contained a number of compromises, and the three-fifths figure represented one such compromise. Before the changes introduced by the Thirteenth and Fourteenth Amendments, the census was used not only to

---

[7] See, "Why We Ask Questions About Hispanic or Latino Origin" in references.

[8] This continued until 1868 when the 14th Amendment—which gave citizenship to all persons born or naturalized in the U.S.—was ratified. The 1870 Census also shows that no slaves were counted. See "Table 1, Population of the United States" in chapter references.

apportion Congress but also to apportion direct taxes of the states.[9] This was important because the delegates from both the Northern and Southern states wanted to maximize their representation but reduce their tax burdens. The Northern representatives were concerned that the enslaved population would give the South too much representation. The Southern representatives were unwilling to be taxed on their slaves. Following the Civil War, the formerly enslaved and their descendants were counted just as all other residents were counted.

The original racism of the three-fifths rule was echoed in later efforts to inhibit or eliminate voting by Black Americans. Jim Crow laws in the South, with their many measures to suppress voting by Black residents, had the effect of increasing the representation of the Southern states (more population led to more members of Congress) and also the effect of magnifying the political impact of the individual white voter (because so many other residents could not vote). Chapter 4 will address the ways in which political objectives are often met using census data. It is important to note here, however, that the data themselves are neutral, and the fact that voter suppression might be attempted is not a reason to avoid the census.

In fact, there is reason to participate in the census even if local officials seek to disenfranchise residents. When the Constitution was written, the three-fifths compromise balanced a good (representation) against a burden (higher taxes). Today the census data reinforce one good (representation) with another good (access to federal funding). This is why the Census Bureau emphasizes the importance of counting everyone, not just voters.

1.9   Costs of the Census

Another issue looms large in the public imagination: how much? Is the census worth the cost? In 2010 the U.S. Census cost $117 per household, compared with $3.93 per household in 1900 (constant 2010 dollars, National Research Council, 2011). Americans could well wonder why the cost has escalated so much and what they are getting for this investment. The 2020 Census is estimated to cost taxpayers some $15.6 billion dollars. The census costs a lot because the country is large and because the Census Bureau works so hard to be sure that everyone is included.

Much of the expenditure comes from the efforts to achieve completeness and accuracy, and especially to ensure that the count is complete. These efforts are the field operations, the process of distributing and collecting the census returns and also in improving information technology and the ability to process the information. The census requires years of planning and testing, and the testing is also expensive.

---

[9] While the direct tax clause appeared to be of great concern to the delegates attending the Constitutional. Convention, directly taxing the states proved difficult to do. Johnson (2004) argues that this was because wealth was not geographically distributed in the same way as population was. The direct tax provision was eliminated when the Fourteenth Amendment eliminated the three-fifths provision. See Johnson (2004) in chapter references.

Between 2012 and 2015 there were seven tests. Larger tests came in the remaining years of the decade. The 2016 Census Test in Houston (Harris County) and Los Angeles (Los Angeles County) emphasized testing procedures for non-response follow-up, the relatively expensive procedure that occurs whenever a household fails to respond in timely fashion. The 2017 Census Test contacted 80,000 households, with an oversampling of tribal areas, to test self-response using the internet. The 2018 Census Test was in Providence, RI, and 52.3% of the households self-responded, which exceeded the Census Bureau's goal for the test. About 700 temporary workers were hired to do the follow-up work in Providence. This rehearsal tested the iPhones that the census takers will use, and it found a considerable improvement in efficiency over the 2010 Census. The 2019 Census Test was a nationally representative test of about 480,000 households.

These dress-rehearsal censuses offer opportunities to pre-test questions and procedures. Because the 2020 Census will be the most technologically advanced census ever deployed, a great deal of the early testing has emphasized the use of the internet for responding, cloud techniques for data storage, and handheld devices for use by employees.

About six months before the census is taken, the Census Bureau conducts a canvassing operation to ensure that its address list of housing is as complete as possible. In the canvassing operation that began in the fall of 2019, satellite imagery was used to identify stability or change in housing and to compare the numbers of visible housing units with the existing census files. This work is supplemented by in-field address canvassing, with listers going door-to-door to confirm and update addresses. Tribal, state, and local governments are encouraged to review these lists to make any needed additions, and the New Construction Program is available for local governments to provide addresses for new units that could be occupied by April 1, 2020.

When there is no response from a housing unit, census personnel will make repeated efforts to secure a completed form or to confirm that the housing unit is vacant. The most expensive way of securing a completed return is to send a trained census taker to the house, which sometimes happens several times. After failing to receive any response, the census taker may speak with neighbors to learn whether the unit is occupied and if so, how many people live there with whatever information can be gathered, such as sex and approximate ages. This information is only as good as the neighbors' information, but it is the last resort as a way to get information. After the field operations are completed, there are some months of data processing, editing, and data analysis, followed by digital and print publications.

Every 10 years the Census Bureau must ramp up its employee base, hiring thousands of temporary workers. An estimated 500,000 workers will be hired for the 2020 Census, making it the largest peacetime mobilization of the country. Accordingly, the costs of recruitment, payroll, training, and supervision will be substantial.

The census is heralded by extensive publicity, with Public Service Announcements released as early as November 2019. An extensive marketing and publicity campaign precedes the enumeration so that as many people as possible are aware that

Census Day is coming. This marketing and publicity campaign, despite the contributions of airtime and media space, represents a considerable expenditure. The census information is mailed out to most households using the U.S. Postal Service. The 2020 Census will encourage online responses, but responses may also be made by phone and by mail.

On top of the personnel and publicity expenditures, there are the costs of data collection and management, data analysis, and a substantial, free publication program. The anonymized data as well as the prepared analyses are made available to the public, with special resources devoted to helping students, teachers, the media, and elected officials understand the results. On the other hand, the census is a unique, significant product for the country, and skimping on the census budget is a false economy.

1.10   Summary

Over time the United States has evolved from an agrarian economy to an industrial economy, to a service economy. In the future, it seems likely that information (or data) in its many manifestations will be the impetus for many new industries, inventions, occupations, and jobs. The U.S. Census, mandated so long ago when the Constitution was first adopted, has served as a cornerstone of the information infrastructure. Along with the other reasons that the census is valuable, it is the only data set that maps every inhabitant of the United States into a geographic area. And unlike many forms of information, the census remains free and public. In addition to its considerable purposes in furthering democracy, the census also provides critical information used in every geographic area for planning, assessment, and decision-making. Its uses are both in the public sector and in the private sector.

References

Anderson M (2015) The American census: a social history, 2nd edn. Yale University Press, New Haven, CT. p. 204, 218, 253, 255, 259–260

Felony Disenfranchisement Laws in the United States (2014). https://www.sentencingproject.org/publications/felony-disenfranchisement-laws-in-the-united-states/. Accessed 24 Nov 2019

Fontenot AE Jr (2018) 2020 Census Non-English Language Report. (2020 Census Program Memorandum Series: 2018.06) (U.S. Census Bureau: 2018). p. 3. https://www2.census.gov/programs-surveys/decennial/2020/program-management/memo-series/2020-memo-2018_06.pdf. Accessed 24 Nov 2019

Groves R (2010a) Quality in a census, some overview thoughts. https://www.census.gov/newsroom/blogs/director/2010/09/quality-in-a-census-some-overview-thoughts.html. Accessed 18 Nov 2019

Groves R (2010b) Repairing a problem: connecting with local leaders. https://www.census.gov/newsroom/blogs/director/2010/07/repairing-a-problem-connecting-with-local-leaders.html. Accessed 18 Nov 2019

Groves R 2010c) Quality assurance and the 2010 census. https://www.census.gov/newsroom/blogs/director/2010/06/quality-assurance-and-the-2010-census.html. Accessed 18 Nov 2019

Groves R (2010d) Why am i being contacted by the Census Bureau—I Returned my Form! https://www.census.gov/newsroom/blogs/director/2010/05/why-am-i-being-contacted-by-the-census-bureau-i-returned-my-form.html. Accessed 18 Nov 2019

Groves R (2010e) A note to my 600,000 new colleagues. https://www.census.gov/newsroom/blogs/director/2010/05/a-note-to-my-600000-new-colleagues.html. Accessed 18 Nov 2019

Groves R (2010f) The clock is ticking. https://www.census.gov/newsroom/blogs/director/2010/04/the-clock-is-ticking.html. Accessed 18 Nov 2019

Johnson CH (2004) Fixing the constitutional absurdity of the apportionment of direct tax. Constitutional Commentary

Levitt J (2019) Nonsensus: pretext and the decennial enumeration. 3 ACS Sup. Ct. Rev. 59. https://ssrn.com/abstract=3469935. Accessed 4 Dec 2019

National Research Council (2011) Change and the 2020 Census: Not Whether But How. Washington, DC: The National Academies Press. https://doi.org/10.17226/13135

Reamer A (2019) The distribution of census-guided Federal Funds to U.S. Communities: five program examples. George Washington Institute of Public Policy

Shryock HS, Siegel JS (eds) (1976) The methods and materials of demography. Academic Press, Inc., San Diego, CA

Snipp MC (2003) Racial measurement in the american census: past practices and implications for the future. Annu Rev Soc 29:575

Stackhouse HF, Brady S (2003) Census 2000 mail return rates. p. v, 1.

Sullivan T (2020) Coming to our census: how social statistics underpin our democracy (and Republic). Harvard Data Sci Rev 3:1 forthcoming.

The U.S. Census Bureau (2019) What is the 2020 census? https://2020census.gov/en/what-is-2020-census.html. Accessed 18 Nov 2019

The U.S. Census Bureau (n.d.-a) 1790 Census population results. https://www.census.gov/library/photos/1790_0052.html. Accessed 24 Nov 2019

The U.S. Census Bureau (n.d.-b) Table I: population of the United States (By States and Territories) in the aggregate and as white, colored, free colored, slave, Chinese, and Indian at Each Census, p. 7. https://www2.census.gov/library/publications/decennial/1870/population/1870a-04.pdf?#. Accessed 1 Dec

The U.S. Census Bureau (n.d.-c) Why we ask questions about hispanic or Latino origin. https://www.census.gov/acs/www/about/why-we-ask-each-question/ethnicity/. Accessed 25 Nov 2019

U.S. Code Title 13 (n.d.) Chapter 1, Section 2. p. 2–3. https://www.govinfo.gov/content/pkg/USCODE-2007-title13/pdf/USCODE-2007-title13.pdf. Accessed 24 Nov 2019

United Nations Department of Economic and Social Affairs, Statistics Division (2008a) Principles and recommendations for population and housing censuses (Statistical Papers, Series M No. 67, Rev. 2). United Nations, New York, pp 5–6

United Nations Department of Economic and Social Affairs, Statistics Division (2008b) Principles and recommendations for population and housing censuses. (Statistical Papers, Series M No. 67, Rev. 2). United Nations, New York. p. 48, 76

University of Virginia Weldon Cooper Center, Demographics Research Group (2018) National population projections. https://demographics.coopercenter.org/national-population-projections. Accessed 24 Nov 2019

Chapter 2
Who, What, When, and Where of the Census

Abstract The objective of the census is to count every resident of the United States once, only once, and in the right place. Although there have been arguments that illegally present immigrants should not be counted, the current interpretation of the Constitution requires everyone to be counted for purposes of representation. This chapter identifies four hard-to-count populations and the census strategies for addressing these populations. Over half a million temporary workers will be trained and deployed, from remote villages in Alaska to the largest cities. Every household in 2020 will receive a short census form that asks only seven questions, but there are controversies about the questions asked, especially the questions on Hispanic origin and race. The census enumeration, because it locates every individual in a specific geographic location, allows the aggregation of census data into towns, cities, counties, school districts, ZIP codes, and other geographic areas of interest to government, business, and citizens.

Keywords Census · Illegal aliens · Hard-to-count populations · Homeless · Hispanic origin · Racial categories

T. A. Sullivan, *Census 2020*, https://doi.org/10.1007/978-3-030-40578-6_2

Representatives shall be apportioned among the several States according to their respective
numbers, counting the whole number of persons in each State, excluding Indians not taxed.
(U.S. Constitution, Amendment XIV)[1]

The U.S. Constitution calls for an enumeration of the population, and it also
specifies one exception: Indians not taxed. The fact that the Founders considered
and included an exception is used as legal precedent for counting *all* people within
the United States. On occasion a legal challenge has been raised to prevent counting
illegally present aliens or even all non-citizens. In every census since 1790, however, non-citizens have been included in the enumeration. The Constitution could be
amended to add exclusions to the count, but for now the enumeration extends to
everyone within the borders of the United States.

The easiest Americans to count have usually been English-speaking homeowners
who have mail delivery and who live in urban or suburban areas. It has been relatively inexpensive to reach these households. In a changing, complex, and diverse
society, however, many household situations are more challenging and expensive to
enumerate. The Census Bureau has developed its strategies for contacting households based on a decade's experience with the American Community Survey (ACS),
which the Bureau also conducts.

2.1 How People Are Counted

The first censuses were conducted by 650 U.S. marshals, but over time trained enumerators conducted the census. As household work patterns changed, and more
women joined the paid labor force, enumerators were less likely to find someone at
home during the day. In addition, the costs of deploying so many enumerators were
rising, and there was a search for a more efficient but reliable means of census-
taking. The U.S. Census Bureau began using the U.S. mail to send census questionnaires with the 1960 Census. In 1970 urban and suburban residents were asked to
use the mail to return their questionnaire. By 1980 around 95% of households used
the mail to return the census questionnaire. The 2020 Census marks the first time
that an online response mode will be encouraged. The advantages to the Bureau of
the online response mode are the easy availability of multiple language forms and
the ability to begin immediately tabulating results. Online responses also allow
simplifying of other field operations; for example, duplicate returns can be more
easily identified.

Beginning in mid-March 2020, about 95% of all households will receive a letter
asking them to complete the census questionnaire online, but also providing them
with information about answering by telephone. Approximately 21.8% of all

---

[1] The Fourteenth Amendment amended Article 1, section 3 of the Constitution by eliminating the
three-fifths provision for counting enslaved persons. This amendment also eliminated the direct
taxation purpose for the enumeration.

households, or about one in every five households, will also receive a paper questionnaire along with this introductory letter. Paper questionnaires will be distributed to areas where the Census Bureau believes that online response will be less likely, although these households will have the options of online and telephone response in addition to the paper questionnaire. The determination of households less likely to respond online is based on ACS results and characteristics of the area that predict lower access to the internet. These characteristics include a higher population over the age of 65, a low rate of internet subscriptions, and a low self-response rate in the ACS.

In 40 of the 50 states, there are geographic areas that will receive English/Spanish bilingual invitations. These invitations will go to about 9.3% of all households based on census tracts in which at least 20% of the households need Spanish language assistance. This determination is based on ACS data for households with at least one person aged 15 or older who speaks Spanish and also reports not speaking English very well. In areas expected to have a low online response, these bilingual invitations will be accompanied by a Spanish/English bilingual questionnaire. In addition to assistance in Spanish, every invitation will contain directions for completing the census questionnaire in eleven additional languages other than English.

Households that have not responded by mid-April will receive a paper questionnaire in the mail. About 5% of all households will have a paper questionnaire dropped off at their home. And less than one percent of the households are expected to be counted in person instead of invited to respond on their own. These contact strategies have been outlined by the Census Bureau with an interactive map that is available at https://gis-portal.data.census.gov/arcgis/apps/webappviewer/index.html?id=7ef5c37c68a64ef3b2f1b17eb9287427 (The U.S. Census Bureau n.d.-a). The use of the ACS to improve census field operations is a good example of how the Census Bureau works between censuses to improve its methods.

2.2 Who Does the Counting?

The Census Bureau employs a professional staff of some 4,000 employees, including statisticians, demographers, information specialists, economists, and other specialists. They work year-round, many of them at the Census Bureau headquarters in Suitland, Maryland. These professionals collect a wide variety of information besides the decennial population of population and housing. They also conduct economic censuses, conduct hundreds of surveys, and collect and analyze many types of data from other agencies. They oversee the ramping up of the census labor force to conduct the census, and they must hire many more temporary workers to complete the census.

The Census Bureau seeks to hire people who live in the communities where they will work (The U.S. Census Bureau n.d.-b).[2] Beginning in September 2019, the Census began recruiting efforts for hundreds of thousands of census takers, thousands of field supervisors, thousands of recruiting assistants, thousands of clerks, and thousands of office operations supervisors (The U.S. Census Bureau n.d.-g).[3] These jobs are temporary, with most of them ending by July 2020. The jobs require flexible hours including nights and weekends to contact people when they are at home. All of the jobs require paid training. Most workers must also have their own transportation and access to a computer and the internet. The task of recruiting, training, and mobilizing this small army of temporary workers is one of the most daunting features of census taking, especially at a time when unemployment is at historic lows.

Using individual enumerators to contact households is the most expensive form of collecting the information. These personal contacts are, however, essential to completing the census, and in particular for non-response follow-up and for counting the hard-to-count population.

2.3   The Hard-to-Count Population

The constitutional mandate, and the good functioning of democracy, require extra efforts to ensure that everyone is counted. The hard-to-count (HTC) population is important to enumerate for purposes of completeness. In a democracy everyone deserves participation in the census and representation in government, and so the effort to find and enumerate everyone is worthwhile even though it is costly. To understand who the people of the United States are, it is important to represent people of all categories as completely as possible. Some of these people are marginalized and have fewer resources, such as migrant farmworkers. Others are sufficiently wealthy that they can isolate themselves in gated communities and or in highrise skyscrapers. People from all parts of the social landscape need to be included for a full picture of who we are.

There are four main categories of hard-to-count people (Chapin 2019).

2.3.1   The Hard To Locate

This group includes travelers and recent migrants, short-term renters, transients and the homeless, and nomadic peoples. In rural areas, homes may be widely dispersed across large areas; in dense urban areas, housing may be new, recently subdivided, or otherwise overlooked. Seemingly vacant units may in fact be inhabited.

---

[2] See, "2020 Census Jobs" in references.

[3] See, "Job Descriptions" in references.

2.3.2   The Hard To Contact

This group can be difficult to reach because of where their homes are. At one end of the economic spectrum, their homes are guarded, such as gated communities. At the other end of the economic spectrum they may have no permanent address, such as the homeless population. They may also be unable or unwilling to be contacted by U.S. mail, telephone, or email.

2.3.3   The Hard To Persuade

These are the people who are reluctant to be enumerated. They might be people who mistrust the government or who are fearful because they are engaged in illegal activity or live in illegal housing. People may not wish to take the time to complete the census because of business, illness, or personal concerns.

2.3.4   The Hard To Interview

Even when people have been located, contacted, and persuaded, the interview process may be difficult because of language barriers, disability or illness, cultural sensitivities, or other reasons.

The Census Bureau has also identified four strategies that can be used to help overcome the challenges posed by the Hard-to-Count population. The first is careful **staff training**, which builds on the experience of previous field workers who have identified and overcome difficult situations. This training is based upon previous studies of the hard-to-count population, demographic analyses of the undercount, and analysis of changes since the previous census. Providing enumerators with practice dealing with simulated difficult situations and providing them with an extensive handbook can help prepare them. Enumerators should also have ready access to supervisors, translators, and others whose assistance might be needed.

The second strategy is **community partnerships**, usually pursued well in advance of the census. These partnerships include local elected officials, community leaders from non-government organizations such as churches and civic groups, and health and education leaders. The Boys and Girls Clubs, for example, are Census Bureau partners for the 2020 Census. Local surrogates can prepare public service announcements, speak at community gatherings, and otherwise rally support for participating in the count. It is especially useful to partner with members of HTC populations who can provide insight and even accompany enumerators, perhaps acting as translators or explaining cultural sensitivities.

For the 2020 Census, the Complete Count Committees represent a sustained partnership between the Census Bureau and local communities to provide a complete count. The local community benefits by getting fair representation and its fair share of federal revenue-sharing based on a more complete census count, and the Census Bureau benefits by having knowledgeable local informants and allies. The work of the committees includes training about the census, planning for community-wide mobilization, boosting and aiding the Census Bureau's advertising and marketing, and encouraging early response by mail, telephone, or email. Members of the committee might make public service announcements or speak at local rallies to encourage early response. For households that do not respond early, the committees explain the role of census takers and the importance of cooperating with them.

The third strategy is **design and accommodations**. These are minor variations in the usual procedures. The 2020 census return will be available in 13 languages, which is an accommodation for those who do not read English or speak English in the case of a visit by a census worker. Alternative methods of completing the census will be available for those who are blind or vision-impaired, or who have a disability that makes responding difficult. For people with limited connectivity, a partnership with a local school or library might provide the needed access (Center for Urban Research 2018).

The fourth strategy is **special programs.** These are major departures from the usual census-taking methods that are needed because of special needs of the HTC population. With the homeless, for example, enumerators may go to locations that provide services such as shelters and soup kitchens. If there has been a natural disaster that has dislocated residents, it may be necessary to pursue special programs with shelters or temporary relocation areas. Seafarers and fishers at sea may have a special enumeration because their residence is in ship's quarters with no predetermined location on census day.

Remote Alaska poses a special challenge for enumeration. There are villages that are accessible only seasonally, and then by dogsled or small planes. Because of the weather, the actual count occurs on earlier dates and village leaders are called upon to help ensure that the count is complete.

2.4   Counting the Homeless

Counting the homeless is a particularly challenging task, and in 2020 the Census Bureau will conduct this count over a three-day period. On March 30, 2020, there will be a count of people who are in shelters. On March 31, 2020, there will be a count of people at soup kitchens and mobile food vans, taking care that anyone counted at a shelter the previous day is not again counted. On April 1, 2020, there will be a count of people in non-sheltered, outdoor locations. This will include people who live on the streets and in cars, and people who are living in informal encampments (U.S. Census Bureau n.d.-f).[4]

---

[4] See, "How We Count People Experiencing Homelessness" in references.

2.5   Counting People In Transitory Locations

The Census Bureau also devotes special days to counting people who live in transitory locations, defined as "campgrounds, RV parks, marinas, hotels, motels, racetracks, circuses, and carnivals" (U.S. Census Bureau n.d.-e).[5] These are locations where people do not normally live year-round. An RV, tent, or houseboat is not counted as a housing unit if all the people staying there report that they have a usual residence elsewhere; in this case, they are reported back at that home. During the period from April 9 to May 4, census takers will visit each location that has been identified as a transitory location, and the census takers will interview the people there using paper questionnaires (U.S. Census Bureau n.d.-c).[6]

2.6   Foreign Countries and Foreign Residence

One issue that faces every census in every country is accounting for one's nationals who are abroad and deciding whether to count foreign nationals within one's own country. Americans who are studying in another country are not included, but foreign students studying in the United States are counted (U.S. Census Bureau n.d.-h).[7]

Foreign visitors and tourists to the United States are not counted. Foreigners who are living in the United States—defined by usual place of residence—are counted, including members of foreign diplomatic corps. There is certainly a line-drawing problem here: for example, if a company sends a foreign national to work in their American subsidiary for six months, then regardless of visa status that living situation seems to describe "usual place of residence."

2.7   *De Jure* Versus *De Facto* Counting

The issue with foreign nationals is a special case that calls attention to the differences between *de jure* versus *de facto* counting. *De jure* counting is counting a person at their place of usual residence, even if the person is temporarily working elsewhere on census day. *De facto* counting means counting these persons wherever they actually are on census day. The Census Bureau says that its objective is to count people "where they usually reside." Therefore, a large hotel hosting a convention over the period March 31-April 2 does not automatically add to the count for the

---

[5] See, "How are people counted at RV Parks, Campgrounds, and Other Transitory Locations?" in references.

[6] See, "Enumeration at Transitory Locations," in references.

[7] See, "Who to Count" in references.

host city. But in practice, the Census Bureau uses a mixture of *de jure* and *de facto* criteria and seeks to avoid duplicate counting.

This issue comes up regularly with college students who live at their colleges. In this case, the Census Bureau counts the students at their usual place of residence, which during the academic year is the residence hall or apartment where the college is located. The students also maintain with the college a permanent address, which is usually where the students' parents live, often in another county or another state. For purposes of the count, however, the permanent address is not the correct address to use in the census.

College towns are eager to have these students counted at their schools, arguing that the college towns provide services for the students (water, sanitation, streets, etc.) for at least nine months of the year. The hometowns are also eager to have these students counted at their permanent residence. So far, however, the Census Bureau continues to count students in their college towns. By contrast, students in boarding schools are counted at their parents' residence, and not at the school.

People who are hospitalized on April 1 are generally counted at their homes, with a few exceptions. Two exceptions are psychiatric facilities and nursing facilities; their residents are counted at the facility. Newborn babies who are still in the hospital are counted at the home where they will live, and the Census Bureau has had a special task force on counting babies and children, because they are often undercounted in censuses. Members of the military deployed abroad are counted at their usual residence in the United States, and on-base housing units are treated just as all other housing units are. Inmates in correctional facilities, such as jails and prisons, are counted at the facility.

There are a variety of other housing issues that confuse the issue of *de jure* and *de facto* counting. April 1 may be too soon for all the snowbirds to have left their southern homes for their northern homes, and so there may be an issue of where they reside "most of the time." The census return includes a mention about whether a person has a second or seasonal home precisely to help determine the correct answer as to where a person should be counted, and to properly account for the vacant second or seasonal home. This circumstance is especially relevant for resort areas that may have many vacancies in April but would be more populous in the winter (for a ski resort) or in the summer (for a beach resort).

2.8 What Questions Are Asked?

Figures 2.1 and 2.2 shows a facsimile census return for the 2020 Census. There are four brief introductory questions, asking for the number of people living in the housing unit on the census date, whether there were any additional people staying there on the census date, whether the unit is owned or rented, and a telephone number for contact. Then for Person One in the unit, the questionnaire asks for name, sex, age, date of birth, Hispanic-Latino-Spanish origin, race (which requires checking a box and then printing origins). For all additional members of the household,

Start here OR go online at [url removed] to complete your 2020 Census questionnaire.
Use a blue or black pen.

Before you answer Question 1, count the people living in this house, apartment, or mobile home using our guidelines.

- Count all people, including babies, who live and sleep here most of the time.
- If no one lives and sleeps at this address most of the time, go online at [url removed] or call the number on page 8.

The census must also include people without a permanent place to live, so:

- If someone who does not have a permanent place to live is staying here on April 1, 2020, count that person.

The Census Bureau also conducts counts in institutions and other places, so:

- Do not count anyone living away from here, either at college or in the Armed Forces.
- Do not count anyone in a nursing home, jail, prison, detention facility, etc., on April 1, 2020.
- Leave these people off your questionnaire, even if they will return to live here after they leave college, the nursing home, the military, jail, etc. Otherwise, they may be counted twice.

1. How many people were living or staying in this house, apartment, or mobile home on April 1, 2020?

   Number of people =

2. Were there any **additional** people staying here on April 1, 2020 that you **did not include** in Question 1?
   Mark ✗ all that apply.
   - ☐ Children, related or unrelated, such as newborn babies, grandchildren, or foster children
   - ☐ Relatives, such as adult children, cousins, or in-laws
   - ☐ Nonrelatives, such as roommates or live-in babysitters
   - ☐ People staying here temporarily
   - ☐ No additional people

3. Is this house, apartment, or mobile home — Mark ☐ ONE box.
   - ☐ Owned by you or someone in this household with a mortgage or loan? Include home equity loans.
   - ☐ Owned by you or someone in this household free and clear (without a mortgage or loan)?
   - ☐ Rented?
   - ☐ Occupied without payment of rent?

4. What is your telephone number?
   We will only contact you if needed for official Census Bureau business.

   Telephone Number

Fig. 2.1 A facsimile of page 1 of the 2020 Census questionnaire. For illustrative purposes only

the same questions are asked with the addition of questions about whether this person usually stays somewhere else (such as a second or seasonal home) and relationship to the first person.

Congress made the decision about what questions would be asked for the early censuses. Since 1950 the Census Bureau has developed a network of advisory committees, hearings, and outreach to communities to develop priorities and rationales for the questions to be asked. This consultative process begins very shortly after one census is completed for the next census. Congressional approval is required for the questions to be asked.

2.9 How Much Information Is Requested?

With every census there has been controversy over asking for minimal information versus seeking more complete information. The 1790 Census was notably short. That first census asked for the number of males above and below the age of 16 (probably related to estimating military manpower), for the number of females, and for the number of enslaved persons (Anderson 2015). When he was a member of Congress discussing the very first census, James Madison argued for much more information, noting that the information would be valuable for understanding the

Person 1

5. Please provide information for each person living here. If there is someone living here who pays the rent or owns this residence, start by listing him or her as Person 1. If the owner or the person who pays the rent does not live here, start by listing any adult living here as Person 1.

What is Person 1's name? *Print name below.*

First Name   MI

Last Name(s)

6. What is Person 1's sex? *Mark* ☐ *ONE box.*

☐ Male · ☐ Female

7. What is Person 1's age and what is Person 1's date of birth? *For babies less than 1 year old, do not write the age in months. Write 0 as the age.*

Print numbers in boxes.

Age on April 1, 2020   Month   Day   Year of birth

☐ years

→ NOTE: Please answer BOTH Question 8 about Hispanic origin and Question 9 about race. For this census, Hispanic origins are not races.

8. Is Person 1 of Hispanic, Latino, or Spanish origin?

☐ **No**, not of Hispanic, Latino, or Spanish origin

☐ Yes, Mexican, Mexican Am., Chicano

☐ Yes, Puerto Rican

☐ Yes, Cuban

☐ Yes, another Hispanic, Latino, or Spanish origin – *Print, for example, Salvadoran, Dominican, Colombian, Guatemalan, Spaniard, Ecuadorian, etc.*

9. What is Person 1's race?

Mark ✗ *one or more boxes AND print origins.*

☐ White – *Print, for example, German, Irish, English, Italian, Lebanese, Egyptian, etc.*

☐ Black or African Am. – *Print, for example, African American, Jamaican, Haitian, Nigerian, Ethiopian, Somali, etc.*

☐ American Indian or Alaska Native – *Print name of enrolled or principal tribe(s), for example, Navajo Nation, Blackfeet Tribe, Mayan, Aztec, Native Village of Barrow Inupiat Traditional Government, Nome Eskimo Community, etc.*

☐ Chinese   ☐ Vietnamese   ☐ Native Hawaiian

☐ Filipino   ☐ Korean   ☐ Samoan

☐ Asian Indian   ☐ Japanese   ☐ Chamorro

☐ Other Asian – *Print, for example, Pakistani, Cambodian, Hmong, etc.*   ☐ Other Pacific Islander – *Print, for example, Tongan, Fijian, Marshallese, etc.*

☐ Some other race – *Print race or origin.*

→ If more people were counted in Question 1 on the front page, continue with Person 2 on the next page.

Fig. 2.2 A facsimile of page 2 of the 2020 Census questionnaire. For illustrative purposes only

economic situation of the new republic. Over time, census enumerators began to ask more detailed information for every member of the household.

The relatively short questionnaire in Figs. 2.1 and 2.2 is the census return for every housing unit in 2020. Between 1940 and 2000, there was a short form similar to the 2020 form, and a much longer form that went to a fraction of the population. The long form asked many questions, some of which caused controversy in the general public, including many items about the physical condition and value of the housing unit, and detailed information about the education, labor force status, occupation, and earnings of the people in the housing unit.

Members of the general public, pundits, and members of Congress argued that these questions were intrusive and unnecessary. The long form took a considerable amount of time to complete, which led to the additional argument that the long form was leading to more non-response and eventually more undercount. The Census Bureau's response—that each question had previously been justified and was needed for a legitimate government purpose—did not satisfy the critics. For the 2010 Census, the long form was eliminated and the monthly American Community Survey, which had begun in 2005, was used instead to provide more detailed information, including information on American cities and towns. The survey had the advantage of providing data throughout the decade, and because it was based on a relatively manageable sample size of 250,000, interviewers were able to persuade respondents to complete the survey.

The United Nations (2008) recommends a great deal of content for the census to collect, far more than the 2020 Census form. The American Community Survey and some other periodic surveys, such as the Current Population Survey, provide much of the detailed information that the United Nations recommends. One item that the United Nations recommends but which the Census Bureau does not ask is religion. One Census Bureau survey in 1957 did ask a question on religion, but the Census Bureau was taken to court over whether the question represented an establishment of religion, something forbidden to Congress under the First Amendment. Since then the Census Bureau has not asked questions about religion.

2.10   Race, Hispanic Origin, and Ancestry

Two questions on the 2020 Census form concern race/ethnicity, and this area is controversial in many quarters. Some commentators have argued that the census should be colorblind, and that it should not be the government's business even to ask people how they identify according to racial category. There is a history of misuse of racial categories. The Nuremberg laws in Germany and Jim Crow laws in the American South both presumed to categorize people based on ancestry, with terrible results for the people who were thus marginalized.

In addition, the term "race" is not scientific. In the early years of the twentieth century, the term "race" was applied in many instances where the twenty-first century might use the term ethnicity, origin, or ancestry. In the 1930 Census, for example, "Mexican" was listed as a racial category. Moreover, for many censuses the information on race was not asked of the respondent but instead the reported information was based upon the visual impression of the interviewer. Perhaps not surprisingly, by the time of the 1960 Census there had begun to be calls to eliminate such questions altogether.

The Civil Rights Movement of the 1960s changed the perceptions of Congress and federal agencies substantially. There came to be renewed interest in the disparities among racial groups, including housing segregation, and these disparities could not be effectively investigated if a racial classification was omitted. By the 1970 Census, the increasing number of residents who were of Hispanic origin—whether from Mexico, Puerto Rico, Cuba, or Spanish-speaking countries of Central and South America—led to development of a question on Hispanic origin that appeared only on the long form. Since the 1980 Census the Hispanic origin question has been asked of every household, on the short form through the 2000 Census and then on the only form in 2010 and 2020.[8]

In 2020, every respondent will be asked to answer question 6 on the census form, "Is this person of Hispanic, Latino, or Spanish origin?" The question is followed by five boxes that one could check: "No, not of Hispanic, Latino, or Spanish origin"; "Yes, Mexican, Mexican Am. [sic], Chicano"; "Yes, Puerto Rican"; "Yes, Cuban"; and "Yes, another Hispanic, Latino, or Spanish origin." Anyone checking this final box will be asked to fill in a specific origin, and some examples are provided: "Salvadoran, Dominican, Colombian, Guatemalan, Spaniard, Ecuadorian, etc."

U.S. immigration law was reformed in 1965, and the reforms took effect in 1968, with the result that the previous limitations on immigration from some parts of the world were removed and family reunification became a more important role in granting visas for entry. For a variety of reasons, including a robust U.S. economy compared with many other countries, immigrants from many different countries entered the United States, and so the census began to reflect a renewed interest in how Americans regarded their ancestry.

This interest is reflected in the open-ended question on the census form that invites people to write in specific origins after indicating race in question 7. For respondents who identify as "White," the form provides examples in parentheses: "German, Irish, English, Italian, Lebanese, Egyptian, etc." For people who say they are "Black or African Am. [sic]" some possible origins are "African American, Jamaican, Haitian, Nigerian, Ethiopian, Somali, etc." For respondents who check the box for "American Indian or Alaska Native," there are instructions to "print name of enrolled or principal tribe(s)." Examples are given.

There is no single Asian or Asian-American box; instead, there are separate boxes for Chinese, Filipino, Asian Indian, Vietnamese, Korean, Japanese, Native Hawaiian, Samoan, Chamorro, and "Other Asian" or "Other Pacific Islander." And then there is a final box for "Some other race—Print race or origin."

---

[8] In the 1950, 1960, and 1970 Censuses, the Census identified Spanish surnames in five Southwestern states. Some other methods that were tried included county and birth and country of parents' birth (1880 to 1970) and Spanish mother tongue (1940 and 1970 Censuses). Jacob S. Siegel and Jeffrey S. Passel, 1979, "Coverage of the Hispanic Population of the United States in the 1970 Census: A Methodological Analysis."

Nothing on the 2020 Census will cause as much effort as tabulating the handwritten entries to the origin items in Question 6 and especially Question 7, and then developing good summaries of the data.

2.11 When Is the Census?

The census date is April 1, 2020, but the actual operational work of begins well before that date. In early 2019, the Census Bureau set up offices in cities around the country. The process of canvassing, or checking every address on every block, required 32,000 workers and ended on October 11, 2019 (The U.S. Census Bureau 2019).[9]

The census enumeration begins in Alaska, where many areas have spotty mail service and villages are isolated by harsh winter conditions for much of the year. The census will begin in Toksook Bay, a village on the Bering Sea that can be reached by land only when the ground is still frozen. When the spring thaw comes, some villagers will leave to fish and hunt. Census takers begin their work on January 21 and enumerate the residents in approximately 240 villages in the most remote parts of Alaska (Lacey 2019). Some remote parts of Maine will also receive early enumeration.

By May 2020, the census takers will begin visiting homes that have not yet responded to be certain that everyone is counted.

U.S. law requires that by December 31, 2020, the Census Bureau must deliver to the President and to Congress the population counts by state that will be needed for apportionment. By March 31, 2021, the Census Bureau will send redistricting counts to the states. This information will be used to redraw legislative districts based on population changes. And through the early 2020s, various census products will be produced to provide further analysis and documentation.

Also of importance, census professionals will begin to analyze errors of coverage and content in the census, and the long planning arc for the 2030 Census will begin.

2.12 Where: The Issue of Census Geography

The census is a unique data source because it associates every resident of the United States with a geographic location. This task is integral to the constitutional purpose of representation. While apportionment requires only the totals for each state, actually drawing the districts requires detail down to small geographic areas. And there is always general interest in aggregating census data with civil boundaries to report totals for towns, cities, and counties.

---

[9] See, "2020 Census In-Field Address Canvassing Operation Ends" in references.

Standard Hierarchy of Census Geographic Entities

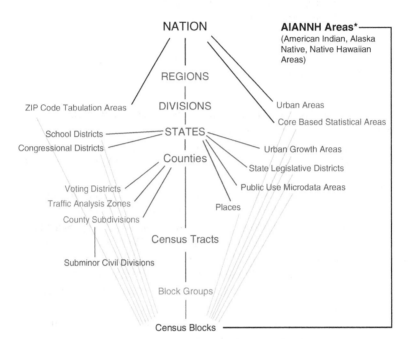

Fig. 2.3 This Census Bureau diagram shows the different geographic entities that can be developed from the census. This includes the smallest geographic level, called the census block. For source see, The U.S. Census Bureau (n.d.-d) "Hierarchy Diagrams"

Figure 2.3 depicts the variety of geographic aggregations that can be developed from the census. The smallest geographic level in the census is the census block. It is rightfully called a "block," because like a building block, it is the foundation for all the other geographic measures. It is not necessarily the same as a city block. Adding together the census blocks develops larger aggregations such as block groups, ZIP codes, school districts, legislative districts, and urban areas.

Block groups can be aggregated to form census tracts, and census tracts can be aggregated to counties. Counties may be subdivided into voting districts and minor civil divisions. Counties can be aggregated into states.[10] The Census Bureau also aggregates states into divisions and divisions into regions. The Census Bureau provides data summaries for additional legal or administrative entities, such as American Indian and Alaskan Native areas and incorporated places.

For larger aggregations, such as counties or cities, there are few concerns about privacy of information. For the smallest geographic areas, however, such as census blocks or block groups, there have been concerns about whether a clever person could reverse engineer the census data to find out information concerning a particular individual. The issue of privacy is addressed further in Chap. 5.

---

[10] Louisiana has parishes.

2.13 Summary

This chapter has covered the basics of the 2020 Census: who is counted and by whom, what questions are asked, when the census operations are conducted, and the geographic reporting of the census. Requiring a massive mobilization of half a million workers and a decade of planning, the U.S. Census is a remarkable undertaking whose results affect the nation in many ways. Because of its significance, the census also arouses controversies over its conduct and its uses. These controversies are discussed in greater detail in the following chapters.

References

Anderson M (2015) The American Census: a social history, 2nd edn. Yale University Press, New Haven, Conn.. p. 204, 218, 253, 255, 259–260

Center for Urban Research, City University of New York Graduate Center (2018) Public libraries and the 2020 Census. https://www.gc.cuny.edu/Page-Elements/Academics-Research-Centers-Initiatives/Centers-and-Institutes/Center-for-Urban-Research/CUR-research-initiatives/Public-Libraries-and-the-2020-Census. Accessed 9 Dec 2019

Chapin MM (2019) 2020 Census: counting everyone once, only once, and in the right place. https://www2.census.gov/programs-surveys/decennial/2020/program-management/pmr-materials/10-19-2018/pmr-hard-to-count-2018-10-19.pdf. Accessed 18 Nov 2019

Lacey ST (2019) Getting to hard-to-reach villages before spring thaw and start of hunting season. U.S. Census Bureau, July 22. 2019. https://www.census.gov/library/stories/2019/07/alaska-remote-areas-always-first-counted-decennial-census.html. Accessed 21 Nov 2019

The U.S. Census Bureau (2019) 2020 Census in-field address canvassing operation ends. https://www.census.gov/newsroom/press-releases/2019/2020-adcan-ends.html. Accessed 21 Nov 2019

The U.S. Census Bureau (n.d.-a) 2020 Census: mail contact strategies viewer. https://gis-portal.data.census.gov/arcgis/apps/webappviewer/index.html?id=7ef5c37c68a64ef3b2f1b17eb9287427. Accessed 21 Nov 2019

The U.S. Census Bureau (n.d.-b) 2020 Census jobs. https://2020census.gov/en/jobs.html. Accessed 21 Nov 2019

The U.S. Census Bureau (n.d.-c) Enumeration at transitory locations. https://2020census.gov/en/conducting-the-count/tl/etl.html. Accessed 21 Nov 2019

The U.S. Census Bureau (n.d.-d) Hierarchy diagrams. https://www.census.gov/programs-surveys/geography/guidance/hierarchy.html. Accessed 20 Dec 2019

The U.S. Census Bureau (n.d.-e) How are people counted at RV Parks, Campgrounds, and other transitory locations? https://2020census.gov/en/conducting-the-count/tl.html. Accessed 21 Nov 2019

The U.S. Census Bureau (n.d.-f) How we count people experiencing homelessness. https://2020census.gov/en/what-is-2020-census/focus/people-experiencing-homelessness.html. Accessed 21 Nov 2019

The U.S. Census Bureau (n.d.-g) Job descriptions. https://2020census.gov/en/jobs/job-details/job-descriptions.html. Accessed 21 Nov 2019

The U.S. Census Bureau. (n.d.-h) Who to count. https://2020census.gov/en/who-to-count.html. Accessed 22 Nov 2019

United Nations Department of Economic and Social Affairs, Statistics Division (2008) Principles and recommendations for population and housing censuses. Statistical Papers, Series M No.67, Rev. 2. United Nations, New York

Chapter 3
Who's Missing? Undercounting and Underreporting

Abstract The proportion of residents missed in the census has declined with each recent census, but it is still too high. Through demographic analysis, post-enumeration studies, and matching of census data with administrative records, the Census Bureau estimates the size and characteristics of the missed population. Children, minority groups, renters, and rural residents are among the most under-counted populations. Although the undercount has sparked many lawsuits, the courts have so far been reluctant to order an adjustment to the reapportionment data as a response to undercount. The stakes are high for states and localities. Besides representation in Congress, the distribution of $1.504 trillion dollars in 316 federal funding programs also is determined by census data, much of that funding intended for the groups that are most likely to be missed. These programs fund health, education, nutrition, and other programs—many of which serve children and low-income households. For the 2020 Census, the Census Bureau is encouraging the creation of Complete Count Committees in states and localities to stimulate higher response rates.

Keywords Census · Undercount · Imputation · Children · Minorities · Rural resident · Renters · Complete count committees

… while the Three-Fifths compromise is a thing of the past, African Americans and other populations of color still suffer from gross undercounts. What exactly does an undercount mean for these communities? It means that they will not be allocated the proper funding for schools, community centers, social services, public roads, transportation etc.—NAACP, "What to Know About the Census" 2019

Thomas Jefferson, then Secretary of State, was in charge of the 1790 Census,[1] and both he and President George Washington believed that the census had missed many inhabitants of the young United States (Anderson 2015 p. 15). Undercounting is nearly inevitable with a large, diverse population that is scattered over a continent and some of whose inhabitants may harbor resentment, suspicion, or just indifference to Washington, D.C. Census professionals, however, have been relentless in their efforts to identify the size and the cause of undercounts, and then to develop field techniques to overcome the obstacles to a complete count.

The problem with undercounts, as the quotation from the NAACP at the opening of the chapter suggests, is that errors of coverage are not randomly distributed across the population. If they were, it would be much easier to apply a statistical remedy. Instead, the evidence over more than seven decades of census taking is that the undercount errors are not random, and the errors disproportionately affect some population groups and some geographic areas.[2] The consequences can be large. The NAACP alludes to the loss of federal funding for communities that are undercounted. They also lose representation, and in the case of reapportionment, a differential undercount could potentially spell the difference between gaining or losing a member of Congress.

In the 2000 Census, the only Americans residing abroad who were counted were federal employees. More than 11,000 Mormon missionaries from Utah who were working abroad were not counted. In the reapportionment after 2000, Utah missed gaining the final seat in the House of Representatives by 856 people. The final seat went to North Carolina instead (Janofsky 2001).[3] Utah sued the federal government, claiming that the count had not been fair. The Utah government officials who were the plaintiffs decried the Census procedure saying, "Such arbitrary, irrational line-drawing cannot be reconciled."

The Census Bureau justified its decision by saying that it had accurate data on federal employees abroad because payroll records and other records could be reliably obtained from federal agencies and the branches of the military. Finding private employees abroad, however, would be difficult and could result in greater

---

[1] The 1790 Census measured the population of the original 13 states, but also included the districts of Maine, Kentucky, Vermont, and the Southwest Territory (which would later become Tennessee in 1796). For more see, The U.S. Census Bureau (n.d.-a, b, c) 1790 Overview.

[2] Hauser, P. (1981). "The Census of 1980." Hauser notes that the Census Bureau program of analyzing the undercount began with the 1950 Census. The concern with undercount goes back to 1790.

[3] After the 1990 census, there were 170 lawsuits filed against the Census Bureau, and many of them alleged undercounts. Utah lost this issue at the level of the district court, but a related undercount issue went to the Supreme Court. *Utah v. Evans*, 143 F.Supp.2d 1290 (2001), D. Utah.

error—of undercount. Utah's response was that the Church of Latter-Day Saints had detailed information on its missionaries abroad and the church would share it with the government.

Utah lost the argument concerning the counting of the missionaries, but the case proceeded to the Supreme Court based on another argument, a statistical technique that the Census Bureau uses to try to correct for undercount. As we shall see, both the undercount and the measures used to correct for it are controversial.

3.1   Who Is Undercounted?

Knowing that there is an undercount raises the interesting issue of how, if people are not counted, can we know who they are? Demographers have a variety of tools to determine whether an undercount occurred, and if so, which people were most likely to be undercounted. Three important methods are reported in the literature. The post-enumeration survey, described in Chap. 1, is a way to check whether the members of a random sample of the population were also included in the census. A second method is to compare questionnaire data from people who participated in census surveys with their return from the census itself. This method might show that some people were in the survey but not the census; it also provides a way to check the reliability of the reported data for the two sources. A third method, called demographic analysis, is to compare the census data with the count that would be expected based on other sources of data, such as birth and death records.

The United States census has been improving over time in the completeness of its count. For the country as a whole, the Census Bureau estimated a net undercount of 1.61% in 1990, a net overcount of 0.49% in 2000, and a net overcount of 0.01% in 2010. The 2010 figure was not significantly different from zero (U.S. Census Bureau 2012).[4] An overcount results from duplications. The small net coverage error was an indicator of a technically sound census

The figure for the entire population, however, conceals the fact that there are people who are still undercounted. Renters, for example, were undercounted in 2010 by 1.1%. The Black population was undercounted by 2.1%. Based on previous censuses, the people most likely to be undercounted have had lower education and lower earnings, have been renters, and have been disproportionately from minority groups. The undercount is larger in rural areas and in central cities. The age group that is most likely to be missed is young children under the age of five.

Given the population changes that occurred in the United States after 2010, there has been an increase in groups for which an undercount has traditionally been a problem. Without a solution, these changes alone would argue for an undercount in the 2020 Census. Researchers at the Urban Institute estimate that demographic

---

[4] See, "Census Bureau Releases Estimates of Undercount and Overcount in the 2010 Census" in references.

shifts alone could lead to an undercount of 0.27% more compared with the 2010 census (Elliot et al. 2019). A coverage error of one-quarter of one percent does not sound big, but it translates into an undercount of nearly a million people.

3.1.1 The Missing Children

Birth and death records provide an independent estimate of the numbers of young children who have been born since the last census and who should be alive to be counted in the current census. These numbers can then be compared with the census counts. This comparison is how demographers know that children under the age of five are the age group that is most likely to be missed. The 2010 Census had a 4.6% net undercount of children under the age of five (Hogan et al. 2013). This represents about one million children who were not counted. An undercount of children is also found in other surveys, such as the American Community Survey, and in the censuses of other countries, so the issue is not limited to the census nor to the United States.

Why are children undercounted? One reason might be that adults completing the census form interpret "person" in the census instructions as meaning "adult." The 2020 Census form specifically instructs the respondents to remember children living in the housing unit, including newborns who might still be in the hospital but who will be coming home to unit's address. The instructions also remind the respondent to consider grandchildren and other children who might be staying in the housing unit.

Another reason that children are often undercounted might be that the young parents are themselves missed, and so are their children. In other words, the entire household is missed from the count (O'Hare 2015). Young parents are likely to be geographically mobile, both because of getting settled into jobs and because of the need for more space when a new child is born into the family. They are also very busy.

Young children are also likely to be missed if they live in complex living situations, such as staying with grandparents or other relatives, or spending time living in different households. Non-Hispanic Blacks, Hispanic, and low-income households have higher net undercoverage of young children than do other racial/ethnic groups (Walejko et al. 2019).

Missing one million children in a population of over 330 million might not sound significant. Its significance is magnified, however, because it points to systematic undercounting of important subpopulations. Subpopulations that are young or have relatively high fertility are thus more likely to be undercounted. If these subpopulations are residentially clustered, the geographic area in which they reside will also be undercounted. Moreover, an undercounted neighborhood will receive fewer of the federal dollars to provide services its residents' need. For school districts, the undercount of young children means that predicting the number of children entering school is harder, and so planning for teachers and physical resources is less efficient.

The Census Bureau developed a task force to consider how to reduce the undercount of young children (The U.S. Census Bureau).[5] In surveys with young families, the Census Bureau found that households with a young child were in general less likely to expect to complete the census form than households without a young child. Whether the 2020 emphasis on internet response will help improve their response rate is hard to assess. On the one hand, 98% of households with a child had access to the internet, especially through smartphones. On the other hand, internet use is lower for non-Hispanic Black, Hispanic, and low-income respondents with young children, some of the groups most prone to undercount.

In an effort to improve the count of children, the Census Bureau has partnered with organizations that focus on child welfare, such as the Annie Casey Foundation, to communicate about completing the census. The Census Bureau is emphasizing outreach to schools, childcare centers, and other places where small children are found so that caregivers can remind the family members to complete a census form (Jarmin 2018). Survey studies have shown that households with a young child have a low level of information about the census, but they also identified several federal programs as important to them, including day care for children, schools, and job training. Non-Hispanic Black and Hispanic respondents also believe that it is important to have census data to enforce civil rights laws. These are themes that the census communications program could address to help reduce the child undercount.

3.1.2 The Missing Minority Group Members

The 2010 Census missed 16 million people, and it overcounted another 8.5 million people who had duplicate counts.[6] Based upon a post-enumeration survey and other analyses, the people who were missed were disproportionately from minority groups. The largest undercount was 4.88% of Native Americans on reservations. There was a 2.07% undercount of non-Hispanic Blacks and an undercount of 1.54% among Hispanics (Mule 2012). The 2010 undercount for non-Hispanic Blacks and for Hispanics represented statistically significant improvements over the undercounts in the 1990 Census—4.57% and 4.99% respectively. Nevertheless, the differential undercount is cause for concern.

In 2020, for the first time a majority of the children under the age of five will belong to a non-white minority group. Two correlates of undercount could lead to a particular issue with minority children.

---

[5] See, "The Undercount of Young Children" in references. See, also O'Hare, W.P. (2014). "Assessing Net Coverage Error for Young Children in the 2010 U.S. Decennial Census."

[6] The overcounted tended to be white, and especially white women. One source of overcount is people who own several housing units or who move during the time of the census, but who complete a return for each unit. Men, especially younger men, are more likely to be undercounted.

Minority group members may be missed in the census for a variety of reasons. One of these reasons is lack of information about the census and its importance to their communities. The Census Bureau has entered into a number of partnerships with non-governmental organizations such as faith-based groups and community centers, with the objective of transmitting more information through "trusted voices." Such partnerships may also lead to practical advice for the operations staff, such as improved access to the internet, more language facility, and the relative success of different media used in advertising.

It is encouraging that the net undercount for minorities has been improving, and yet the differential undercount is legitimately an issue of concern for the Census Bureau, for elected officials, and for the communities themselves (U.S. Government Accountability Office 2018).

3.1.3 The Missing Renters

About 1.9% of renters were missed in 2010, and there has been an undercount of renters for several consecutive censuses (Mule 2012, p. 17). Multi-unit buildings are considered hard to count, and 61% of renters live in multi-unit buildings. Renters tend to be mobile and therefore harder to locate. They might be moving on or near census date, and whoever next rents the unit is unlikely to have a forwarding address. Short-term renters in particular move frequently. Especially for non-family households, such as three or four roommates sharing an apartment, some roommates may stay for only short periods of time and not be well known to the person completing the form.

People living in poverty are more likely to rent, and their ability to pay their rents on time may be problematic. In 2015 there were 2.7 million renters faced with eviction. Following an eviction, there may be no way to update the address of the household that has left. Renters tend to have lower levels of education than owners, and lower levels of education are also correlated with being undercounted. Housing instability may be a particular problem in central cities where rents are rising.

The undercount of renters is especially concerning because there are proportionately more rental households today, in part because of the collapse of the mortgage markets after the Great Recession that began in 2008. In the largest cities of the United States, the majority of households rent their homes. About 37% of all Americans are renters (Leadership Conference Education Fund 2018).

3.1.4 The Missing Rural Residents

Rural areas tend to be undercounted more than urban areas (Hauser 1981). In Census Bureau usage, a rural area contains fewer than 2500 residents. In the 1990 Census, there was a 5.3% undercount reported for white rural renters

(West and Robinson 1999). For Hispanic rural renters, the undercount was quite large at 15.8% (Sherman 2018). Non-metropolitan households generally have a higher response rate than metropolitan areas, so the principal issue is probably not resistance so much as isolation and low density. Rural areas may have non-conventional housing, such as occupied trailers or recreational vehicles. Many rural homes do not have street addresses, and some rural areas do not have home mail delivery. Instead, they get their mail at the post office. Rural areas are more likely than other places to rely upon the personal visit of a census taker, and the relative difficulty in finding them raises the cost of reducing the undercount.

3.2 Efforts to Correct the Undercount

Reducing the undercount is an obvious operational goal for any census. The Census Bureau, however, received as much as $3.3 billion less than requested for 2020 Census operations. The goal of reducing the differential undercount is in tension with the goal of reducing (or at least containing) the cost of the census. Efforts to reduce the undercount are typically expensive and concentrated in the non-response follow-up after Census Day. In 2010 there were six separate attempts to visit a non-responding housing unit (Keller 2015). Such intensive follow-up efforts are not envisioned for 2020, mostly to avoid the substantial personnel expense.

Even the research on how to improve the count is expensive. Funding cuts have affected some of the research directed at improving the count. Because of a lack of funding, the Census Bureau canceled its main field tests of census methods in rural areas in both the 2017 and 2018 tests. Yet rural areas remain prone to undercount for the reasons discussed above. Concern about the effects of budget on the 2020 Census mission led the Government Accountability Office to put the 2020 Census on its High-Risk List in 2017 (U.S. Government Accountability Office 2017).

The Census Bureau will encourage the use of the internet as its preferred means to answer the 2020 Census. The internet self-response option is efficient and if widely used would reduce costs. Its critics, however, say that the methods are insufficiently researched and there are still too many people who do have access to the internet or who do not wish to use it for reasons of privacy (see Chap. 5). The hard-to-count populations are the least likely to use the internet, and the efforts to reach those populations are most affected by funding cuts. Rural areas are the most likely to be affected by the so-called last mile effect, which is the difficulty of extending broadband to remote areas. Elliott et al. (2019) argue that the emphasis on internet response does little to improve the response from the hard-to-count populations.

3.2.1 Editing and Imputation

One area of research into an improved count is better editing and imputation. In essence, both editing and imputation involve the use of statistical methods to deal with missing information to yield a better count. Editing refers to changing answers to a census form that are internally inconsistent. Imputation refers to filling in the unanswered questions with plausible answers or, in cases where there is no return, coming up with a statistically sound estimate for what the information would be. Banister (1980) finds editing and imputation acceptable if they are part of the actual fieldwork operation, but she cautions against what she calls "semi-informed and blind" imputation because it decreases quality.

There are two types of imputation in the census, count imputation and characteristics imputation (Cohn 2011a). Count imputation is, as its name implies, about imputing people for the count. There are three types of count imputation. Status imputation is deciding whether a structure is also a housing unit. An example might be a rural fishing camp—is it really a housing unit, or is it occupied only occasionally? The second type of count imputation is occupancy imputation, which is deciding whether a housing unit is occupied. Following the 2010 Census, New York City filed a challenge to the census alleging that as many as 50,000 occupied units had been missed.[7]

The third type of count imputation is household-size imputation. If a structure is a housing unit, and if it is determined to be occupied, then how many people live there? About 0.24% of all addresses fell into this category in 2010; this amounted to 325 thousand addresses, from which 1,163,462 people or 0.39% of the total population were added to the count (Cohn 2011b). The state with the largest number of people added to the count was Texas, with 143,813 people added to the count.

Critics argue that this type of imputation amounts to the Census Bureau "making it up." Legal issues have arisen about whether imputation should be included in an "actual enumeration." Returning to the *Utah v. Evans* case, it eventually reached the Supreme Court, but not on the issue of the uncounted Mormon missionaries. Rather it reached the Supreme Court on the issue that imputation had given North Carolina too many "extra" people.[8] In a 5-4 decision, the Supreme Court decided that imputation was a legitimate practice. The Court's reasoning was that imputation is "inference," not sampling (sampling is discussed below).

---

[7] Cohn, D. (2011a, b). "New York City Files Census Challenge." Miami, Washington, D.C., and 45 other cities and counties challenged census counts. These challenges could not affect the census count for apportionment, but were filed in hopes of an adjustment to yield more federal revenue-sharing.

[8] The Census Bureau also uses the term "curbstone" to describe the fraudulent practice of making up fictitious census returns—and from time to time a census taker has been accused of curbstoning. To an observer, the issue might arise as to whether some imputed households were perhaps fictitious. In fact, however, there is considerable staff time devoted to an imputed household, certainly under the operational plan for the 2010 Census.

Characteristics imputation is a statistical method for supplying missing data in a census return. Suppose that a family fails to indicate answers to the Hispanic origin question. Normally there will be a telephone follow-up to get the information, but if follow-up attempts fail, then the information could be inserted into the file through one of several statistical practices. One practice is hot-deck editing. This practice is based on the principle of homogeneity, or the fact that people living in a small area tend to be similar to one another (U.S. Census Bureau).[9] In this practice, the computer will be programmed to find the last complete household record from the same census tract that matches the incomplete household record. Then the incomplete record will be assigned the values that had been recorded in the complete record. Overall, the percentage of cases that receive this type of editing is small, and because it is small it is likely that it has little effect on the uses of the census data. Still, it is possible that this form of editing creates errors that could affect a federal funding formula. For 2020, however, the Census Bureau plans to use a different technique, which is looking for information for the household in other administrative records.

3.2.2   Administrative Records

The Census Bureau has been researching the use of administrative records to provide information from missed households or for missing data on incomplete household records (Keller 2015). The government has multiple databases that contain physical addresses and could be used to provide information on the people who live at that address. Some examples of these data bases are Internal Revenue records, Medicare and Medicaid records, and Social Security records. If there has been no return from a certain address, that address could be searched to compile a household roster. Some administrative records also include information on sex, age, and date of birth, and would permit some estimate of the number of people at the address.

Administrative records form a source of proxy information that might be more accurate than speaking to neighbors or landlords. On the other hand, given the reality of a population with births, deaths, and household moves, the administrative records may also be a bit stale. Still, with funding cutbacks for non-response follow-up in 2020, administrative records are a promising approach.

The criticism of the administrative records approach is principally that it has received too little research, certainly for deployment on the scale that the census will require (Mervis 2019). Researchers from the Urban Institute have argued that the census undercount could rise to 0.84% (still less than one percent) given both the demographic changes and the changes in census procedures (Elliott et al. 2019).

---

[9] See "Missing Data, Edit, and Imputation" in references.

3.2.3   Sampling

The Census Bureau has contributed substantially to the development of sampling theory and techniques. The Census Bureau fielded a national sample survey of unemployment in 1937. In the 1940 Census, a sample of 5% of households were asked supplementary questions. From 1950 until 2000, a sample of residents received a long form for the Census, while most residents answered only a few questions on the short form. By 2005, the American Community Survey was being deployed and its use replaced the long form census questionnaire in the 2010 Census. The Census Bureau also conducts the Current Population Survey, the Survey of Income and Program Participation, and many other surveys.

A good sample is representative of the population from which it is drawn. A sample of your blood, for example, is assumed to be representative of all of your blood whether it was drawn from your right arm or your left arm. One advantage of sampling is that statisticians can mathematically calculate the sampling error. Numbers derived from two different surveys can be examined to see whether the difference between them is "real" or merely due to sampling error—that is, the chance that the sample that was drawn was not representative of the population it was supposed to represent.

Given its expertise with sampling and given the vexatious problem of the under-count, it is not surprising that Census Bureau professionals developed some meth-ods to use sampling to adjust the census figures for undercount. The basic technique was something like the post-enumeration survey. The percentage of people who were in the sample survey but had not been counted in the census could be assumed to be representative of all the people who did not answer the census, and their char-acteristics used to adjust the census data. One flaw in this idea is that there are some people who were not counted in the census, fell into the sample, and then were not counted in the sample.

The Clinton Administration proposed to adjust the data from the 2000 Census in this way, but the Supreme Court ruled against the use of sampling for purposes of adjusting the count for reapportionment (Greenhouse 1999).[10] Sampling, the Court argued, was not an "actual enumeration" of the people. The Court suggested that the Census could adjust the census data for the states to use in redistricting, but the Census Bureau decided not to do this because they believed that the adjustment procedure did not meet their strict internal quality standards.

For the 2020 Census, sampling has been used in testing content and techniques, and it will be used in quality checks following the census. The Census Act of 1954, as amended, gives the Census Bureau authority to use sampling for many purposes. Reapportionment, however, is not one of them.

---

[10] This was a 5-4 decision. See, *Department of Commerce v. United States House (98-404).*

3.2.4   Regression Analysis

Many of the variables that predict undercount are correlated with one another. For example, minority populations are more likely to live in poverty, to be renters, to have lower education, and to have somewhat higher fertility. Alone these variables could predict a greater chance for being undercounted. A multiple regression analysis offers a way to control simultaneously for a number of variables and estimate their relative effects. One proposed effort to use regression analysis as a means to adjust the census for undercount became important in a different court challenge, this one by the State of New York against the Census Bureau.[11] In this challenge, which was ultimately unsuccessful, New York proposed using a regression formula to correct the post-enumeration results (Ericksen and Kadane 1985). This formula was criticized by experts hired by the defendants on a number of technical grounds (Freedman and Navidi 1986). Twenty trial days of testimony by the competing statistical experts did not persuade the Court that this form of adjustment should be deployed, especially for sub-areas.

3.3   The Effect of the Undercount On the States

Given the substantial number of court cases and challenges to the census data, it is apparent that the states are well aware of the potential loss they face if an undercount differentially affects them. The number of representatives in the House is fixed at 435, so one state's gain is literally another state's loss. Moreover, the number of federal dollars available for distribution to the states is in effect another zero-sum situation, in which dollars distributed to one state are lost to every other state.

3.3.1   Loss of Representation

As the State of Utah argued in *Utah v. Evans*, a member of the House of Representatives may be lost through a small difference in state counts. So even if an undercount is relatively small for the entire country, if it is concentrated in relatively few states it could be sufficient to affect the reapportionment. One estimate of the projected effect of undercount shows eight seats in the House in play, with Alabama, California, Illinois, Michigan, New York, Pennsylvania, Rhode Island, and West Virginia each losing one seat. The

---

[11] *Cuomo v. Baldrige*, 674 F. Supp. 1089, 1987, SDNY. In this case, New York argued that the state and New York City had been undercounted in 1980 and asked for a Court order to statistically adjust the 1980 Census. New York argued that the undercount had caused New York to lose one or more seats in Congress.

states that would gain seats are Colorado, Florida (two seats), Montana, North Carolina, Oregon, and Texas (two seats).[12]

3.3.2 Loss of Federal Dollars

Reamer (2019) identifies 316 federal funding programs whose distribution depends at least in part on census data.[13] When there is a differential undercount, a great deal of money is at stake. For example, Title I provides funds for states to distribute to local education agencies. Although there are different formulas for different parts of the program, all formulas rely on a count of poor children aged 5–17 in the education district. An undercount either of the poor or of children, both of which have been documented in previous censuses, would lead to a misallocation of these funds. Reamer estimates that for each poor child missed in the census, a school district would not receive $1695. This is a national average and local numbers would vary depending upon the community.

Another example of the effect of differential undercount is the distribution of money that supports public health, such as Medicaid and the Special Supplemental Nutrition Program for Women, Infants, and Children.[14] As Strane and Griffis (2018) note, the same factors that lead to people being undercounted are also the factors that lead them to face public health concerns.

It is instructive to see how the undercount affects one program. Medicaid is the third largest program in the federal budget. Medicaid is of particular interest to the states because states jointly fund the program with the federal government. The formula used to determine the federal government's share of Medicaid funding is inversely proportional to the state's per capita income. This means that as per capita income rises, the share of Medicaid funding from the federal government falls. If an undercount differentially affects the poor—which seems true—then the per capita income will be artificially increased. Perversely, then, a state with a large undercount of poor people will receive a smaller share of the available dollars. And those missed poor people are the very clients most likely to use Medicaid. The inclusion of the undercounted poor would have reduced the per capita income, providing more funds for Medicaid.

---

[12] McGhee, E., Bohn, S. & Thorman, T. (2019). "The 2020 Census and Political representation in California: technical Appendices." Their estimation assumed both a low accuracy census and an immigration-related undercount. They also show estimates using the Weldon Cooper Center's population estimates, and with these estimates Texas gains three seats, Florida gains one, and California does not lose a seat. Ohio would lose a seat.

[13] Reamer also gives examples of job training and community development programs.

[14] Other programs they discuss include Section 8 housing vouchers, Head Start, Early Head Start, the Low-Income Home Energy Assistance Program, and Title I grants to schools. These programs use different formulas that rely on census data in different ways.

3.4 Complete Count Committees

A differential undercount pits states against one another for representation and for federal funding. Given the states' interests in having a good count, it is not surprising that most states have stepped forward with plans to support and encourage the 2020 Census. The Census Bureau has developed a program for Complete Count Committees. States, tribal governments, and localities are encouraged to develop these committees of elected officials, volunteers, officials from non-profit groups, and others in a position to help improve the response rate (U.S. Census Bureau).[15] The Complete Count Committees use local knowledge to educate their communities, publicize the census, and encourage cooperation of local governments with the Census Bureau.

Perhaps the greatest effort is being put forth by the State of California, which is investing $187 million in the 2020 Census. Twenty-three states have funded complete count efforts, with Illinois allocating $30 million and New York $20 million (National Conference of State Legislatures 2019). New York City has earmarked $40 million for its census efforts. In states where no appropriation has been made for a complete count, such as Oklahoma, Nebraska, and Texas, localities are concerned that their states are being set up for considerable losses of funding.

Texas is slated to gain representation, even if there is a considerable undercount, and the possibility for an undercount in Texas is high. As mentioned previously, Texas was the state that gained the greatest addition to the count from imputation. Roughly 7 million Texans live in hard-to-count areas, and much of the state is rural, with little access to the internet. The substantial federal funding that could be lost has been sufficient to stimulate local Complete Count Committees in Texas, even if so far the state has refused to appropriate funding to participate.

3.5 Summary

As with so many things in life, prevention is the cheapest, best solution to the problem of the undercount. Even though a perfect census is not possible, substantial work can be done in advance of the census to encourage even those in hard-to-count areas to complete their census return. After Census Day, there are measures such as non-response follow-up and imputation that can correct at least some non-response. Minimizing the undercount of children, minorities, renters, rural residents, and those in the hard-to-count populations should be acknowledged as part of the democratic process to ensure that everyone counts.

---

[15] See, "2020 Census Complete Count Committees Guides" in references.

References

Anderson M (2015) The American census: a social history, 2nd edn. Yale University Press, New Haven, CT. pp. 204, 218, 253, 255, 259–260

Banister J (1980) Use and abuse of census editing and imputation. Asia Pac Cens Forum 6(3):16–18,20

Cohn, D. (2011a) Imputation: adding people to the census. Pew Research Center, Social and Demographic Trends. Retrieved from https://www.pewsocialtrends.org/2011/05/04/imputation-adding-people-to-the-census/. Accessed Dec. 9, 2019

Cohn, D. (2011b.). New York City files census challenge. August 10. Pew Research Center, Social and Demographic Trends. Retrieved from https://www.pewsocialtrends.org/2011/08/10/new-york-city-files-census-challenge/. Accessed Dec. 9, 2019

Elliott, D., Santos, R., Martin, S., and Runes, C. (2019) Assessing miscounts in the 2020 census. Urban Institute, June 4, 2019. Retrieved from https://www.urban.org/research/publication/assessing-miscounts-2020-census. Accessed Nov. 23, 2019

Ericksen EP, Kadane JB (1985) Estimating the population in a census year: 1980 and beyond. J Am Stat Assoc 8:98–131

Freedman DA, Navidi WC (1986) Regressions models for adjusting the 1980 census. Stat Sci 1(1):3–39

Greenhouse, L. (1999) Jarring democrats, court rules census must be by actual count. *New York Times* (January 26), A1

Hauser P (1981) The census of 1980. Sci Am 245(5):53–61. Retrieved from www.jstor.org/stable/24964600. Accessed Nov. 23, 2019

Hogan H, Cantwell P, Devine J, Mule V, Velkoff V (2013) Quality and the 2010 Census. Popul Res Policy Rev. https://doi.org/10.1007/s11113-013-9278-5

Janofsky, M. (2001) Utah, in Census war, fights North Carolina for house seat. *New York Times* A:18 (February 8)

Jarmin, R. (2018) Improving our count of young children. July 2, 2018. Retrieved from https://www.census.gov/newsroom/blogs/director/2018/07/improving_our_count.html. Accessed Nov. 26, 2019

Keller, A. (2015). Imputation Research for the 2020 Census. Census Working Papers DSSD-WP2015-3. Retrieved from https://www.census.gov/content/dam/Census/library/working-papers/2015/dec/DSSD-WP2015-03.pdf. Accessed Dec. 9, 2019

Leadership Conference Education Fund. (2018). Will you count? Renters in the 2020 Census. July 1. Retrieved from Civilrightsdocs.info/pdf/census/2020/Renters-HTC.pdf. Accessed Nov. 26, 2019

McGhee, E., Bohn, S. and Thorman, T. (2019) The 2020 Census and Political representation in California: technical appendices. Public Policy Institute of California, Table B4. Retrieved from https://www.ppic.org/publication/the-2020-census-and-political-representation-in-california/. Accessed Dec. 9, 2019

Mervis J (2019) Census citizenship question is dropped, but challenges linger. Science 365(6450):211. https://doi.org/10.1126/science.365.6540.211

Mule, T. (2012) Census coverage measurement estimation report: summary of estimates of coverage for persons in the United States. Retrieved from https://www.census.gov/coverage_measurement/pdfs/g01.pdf. Accessed Nov. 26, 2019

National Conference of State Legislatures. (2019). "2020 Census Resources and Legislation." Retrieved from http://www.ncsl.org/research/redistricting/2020-census-resources-and-legislation.aspx. Accessed Dec. 9, 2019

O'Hare W.P. (2014) Assessing net coverage error for young children in the 2010 U.S. Decennial Census. Center for Survey Measurement Study Series (Survey Methodology #2014-02). U.S. Census Bureau. Retrieved from http://www.census.gov/srd/papers/pdf/ssm2014-02.pdf. Accessed Nov. 21, 2019

O'Hare WP (2015) Potential explanations for the high net undercount of young children in the U.S. Census. In: The undercount of young children in the U.S. Decennial Census. SpringerBriefs in population studies. Springer, Cham. https://doi.org/10.1007/978-3-319-18917-8_7

Reamer, A. (2019) The distribution of census-guided Federal Funds to U.S. Communities: five program examples. George Washington Institute of Public Policy

Sherman, A. (2018) The rural undercount. The Census Project. Retrieved from https://thecensus-project.org/2018/03/05/the-rural-undercount/. Accessed Nov. 21, 2019

Strane D, Griffis HM (2018) Inaccuracies in the 2020 Census enumeration could create a misalignment between states' needs. Am J Public Health 108(10):1330–1333. https://doi.org/10.2105/AJPH.2018.304560

The U.S. Census Bureau. (2012) Census Bureau releases estimates of undercount and overcount in the 2010 Census. Retrieved from https://www.census.gov/newsroom/releases/archives/2010_census/cb12-95.html. Accessed Nov. 26, 2019

The U.S. Census Bureau. (n.d.-a) 2020 Census Complete Count Committees Guides. D-1280(RV). Retrieved from https://www.census.gov/content/dam/Census/newsroom/press-kits/2018/ccc-guide-d-1280.pdf. Accessed Dec. 9, 2019

The U.S. Census Bureau. (n.d.-b) Missing data, edit, and imputation. Retrieved from https://www.census.gov/topics/research/stat-research/expertise/missing-data.html. Accessed Dec. 9, 2019

The U.S. Census Bureau. (n.d.-c) The undercount of young children. Retrieved from https://www.census.gov/programs-surveys/decennial-census/2020-census/research-testing/undercount-of-young-children.html. Accessed Nov. 26, 2019

U.S. Government Accountability Office. (2017). Progress on many high-risk areas, while substantial efforts needed on others. February. GAO-17-317. Retrieved from https://www.gao.gov/assets/690/682765.pdf. Accessed Dec. 9, 2019

U.S. Government Accountability Office. (2018). 2020 Census: actions needed to address challenges to enumerating hard-to-count groups. GAO-18-599. Retrieved from https://www.gao.gov/assets/700/693450.pdf. Accessed Nov. 26, 2019

Utah v. Evans. (2001). 143 F.Supp.2d 1290, D. Utah

Walejko, G., Shia, L., Scheid, S., and Griffin, D. (2019) Researching the Attitudes of Households Reporting Young Children—a Summary of Results from the 2020 Census Barriers, Attitudes, and Motivators Study (CBAMS) Survey. June 10. Retrieved from https://www.census.gov/programs-surveys/decennial-census/2020-census/research-testing/undercount-of-young-children.html. Accessed Nov. 26, 2019

West, K. & Robinson, J.G. (1999) What do we know about the undercount of children? Census Bureau Population Division working paper. (August)

Chapter 4
The Citizenship Issue and Gerrymandering

Abstract The census tells the story of population composition—the country of birth, mother tongue, ancestry, and race of the population. Therefore, the census also tells the story of immigration. Although a question about citizenship had been asked as recently as the 1950 Census, it was then asked only of the foreign-born. When Secretary of Commerce Wilbur L. Ross, Jr., announced that there would be an additional question on the 2020 Census on citizenship, the resulting lawsuits were eventually heard by the Supreme Court. The Court narrowly ruled that the question would not be included on the 2020 Census. An important issue in the public controversy was the fear that a census citizenship question would increase the undercount among immigrants and Hispanics.

In the same judicial season, the Court ruled that gerrymandering (except racial gerrymandering) was a political issue beyond the jurisdiction of the courts. Some political strategists have argued for redistricting the state legislatures using voting-age citizens rather than complete population counts. Such a change is believed to favor older, white, suburban residents who are more likely to vote Republican.

Keywords Immigration citizenship question · Gerrymandering · Redistricting

> Often overlooked in the debate over immigration policy and "sanctuary cities" ...is the real boost that such immigrants give to the political power of these jurisdictions. The gross population, including illegal immigrants, is counted in determining representation in the House and state legislatures, rather than the citizen population. Therefore, the more undocumented residents a jurisdiction has, the more legislative seats it gets, both in Washington and in its state capitol. (Lucas 2018).[1]

For many years the U.S. Census has been an important source of information about the diversity of the population, and this information has inevitably interacted with legislators and litigators concerned about immigration policy. Immigration is the entry of a person born elsewhere into the United States for the purpose of permanent residence. Immigrants are not granted citizenship until they undergo the legal process of naturalization. For the 2020 Census, the most inflamed issue has revolved about whether the census form should include a question about citizenship. The U.S. Supreme Court decided (5–4) in the summer of 2019 that the question should not be asked in 2020, but the reasoning was that Secretary of Commerce Wilbur L. Ross, Jr., did not give the real reason for including the question in the census as required by the Administrative Procedure Act.[2] What that reason might have been, against the backdrop of the census and issues of immigration, will be a principal issue raised in this chapter.

4.1   The Census and Immigration

In colonial times European settlers generally viewed the North American continent as empty and in need of labor. Virginia's (1609) Commonwealth Charter welcomed "any other strangers that will become our loving subjects."[3] Over time, however, a sense grew that Americans were a nation distinct from those they had left behind in Europe. The Constitution shows an awareness of natives differing from immigrants: the president of the United States must be a "natural-born citizen."[4] This was not to say that immigrants were barred from high office, however. Article I stipulated the years of citizenship that were required before a naturalized citizen could be elected to become a senator or a representative.[5]

---

[1] James W. Lucas (2018) "Why we need to count citizens in the 2020 census." Washington Examiner, January 25. Retrieved from https://www.washingtonexaminer.com/why-we-need-to-count-citizens-in-the-2020-census on November 26, 2019.

[2] The Administrative Procedure Act provides for the functioning of executive agencies, and with respect to the Secretary of Commerce and the Director of the Census Bureau, there are specified information postings, reports, and deadlines specified.

[3] Massachusetts's' charter (1629) had similar sentiments.

[4] U.S. Constitution, Article 2, section 1.

[5] Article 1, Section 3, Clause 3 states that one of the qualifications to be a U.S. Senator is that the person must have a U.S. citizen for 9 years. Article 1, Section 2, Clause 2 states that one of the qualifications to be a member of the U.S. House of Representatives is that the person must have been a U.S. citizen for 7 years.

For the census, however, the Constitution made no such distinctions. The Constitution specified that "Indians not taxed" were exempt from enumeration, but no other exclusions were made.[6] By eliminating most of the indigenous population, the 1790 Census was an enumeration of immigrants, both slave and free, and their descendants.

Before 1875, relatively little federal policy attention was paid to immigration (Cafferty et al. 1983). There was action at the state level, however. The midwestern states allowed immigrants to vote, hoping to attract more of them to the interior of the United States. The eastern seaboard states, on the other hand, began forbidding immigrants from voting. The Irish potato famine of the 1840s had changed the character of immigration on the East Coast. The large influx of Irish Catholics aroused political and economic opposition. The political opposition took form in new political parties, the Know-Nothing Party and the Native American Party. These parties demanded long residencies before citizenship and literacy before an immigrant was eligible to vote.

Between 1875 and 1965, immigration policy was predominantly based on discrimination against countries of origin, although there were also acts to exclude undesirable individuals.[7] In the 1880s concern arose on both coasts of the United States about immigration—concern about Chinese and other Asian immigrants to the West Coast, and concern about the "new immigrants" from southern and eastern Europe on the East Coast.

A series of statutes was enacted by Congress to restrict immigration from Asia. Chinese women were explicitly banned from entry. Chinese laborers were first limited and then banned. The Japanese government entered into a "gentleman's agreement" not to issue passports for Japanese citizens to emigrate to the United States. In 1917 Congress created the "Asiatic barred zone," barring both Asians and people of Asian ancestry.[8]

The immigration stream on the East Coast, which had previously been predominantly from the British Isles and northern Europe, had shifted to southern and eastern Europe. Poles, Hungarians, Italians, Greeks, and Russians came to dominate the new East Coast immigrant stream. They did not speak English and often they were not Protestant. Both Catholics and Jews were well represented in this new immigration, and there were concerns that their settlement in the large cities of the East would present culture issues. The Dillingham Commission (1911) concluded that northwestern European immigrants were more desirable ("assimilable") than those from southern and eastern Europe.

---

[6] The "three-fifths" compromise for counting enslaved people is discussed in Chap. 2.

[7] Between 1875 and 1882, Congress barred from entry convicted felons, "lunatics and idiots," and persons likely to become a public charge. By 1891 polygamists and persons with contagious diseases were excluded. In 1903 epileptics and professional beggars were excluded. In 1917 Congress added alcoholics, stowaways, and vagrants (Cafferty et al. 1983:41–2).

[8] Filipinos were not restricted until 1946 because of they were considered U.S. nationals until Philippine independence after World War II. Token visa numbers were also given to China after World War II and token numbers were extended in 1952 to other Asian countries.

Restricting immigration was a contentious issue in Congress during the decade following World War I. Given a great deal of pseudo-scientific commentary about various "races" in Europe, there was strong interest in restricting the new stream of southern and eastern Europeans in favor of more immigrants from the British Isles, northern and western Europe. The mechanism of restriction was the problem, and the eventual solution in the National Origins Act of 1924 was to allocate entry based on the proportion of the U.S. population from a specific country as reported in the 1890 Census. In 1890, the immigrant stream was still heavily from Great Britain, Ireland, Germany, and a few other countries. The large influx of Italians, Poles, and others came after 1890, and so the use of this date was strategic.[9]

The law also numerically restricted the total number of immigrants who would be accepted. The effect of this restriction was to dramatically reduce the available slots for southern and western Europe. This law set the basis for the United States' later reactions either sharply limiting or rejecting refugees and displaced persons from Europe throughout the 1930s and well until after World War II ended. For our purposes, however, the critical issue is that census data became pivotal for another government function, the regulation of immigration.

Meanwhile, along the southern border of the United States there were minimal restrictions. Individuals could be excluded for disease, criminality, and other individual characteristics, and they were expected to pay a head tax. But there was no exclusion of any particular nationality. The Monroe Doctrine had warned the European powers to stay out of the New World. Implicit in this formulation was an identification of the United States with other countries from North, Central, and South America, and the Caribbean. Immigration was open to anyone from the Western Hemisphere—Canada, Mexico, all of Central and South America, and the Caribbean. When Cuban refugees streamed into the United States after the Cuban Revolution in the late 1950s and early 1960s, they were welcomed to settle in the United States.

European immigration streams had settled mostly in the large Northern cities, with some European immigration in a few West Coast ports. The South and Southwest, however, traditionally received fewer immigrants, and business interests there experienced labor shortages. Zolberg (2006) explained that Congress left the back door unlocked for cheap Mexican labor. Given the general ferment in Congress over immigration in the 1920s, there had been concern about concentrations of Mexican Americans in the Southwest (Prewitt 2013, p. 76). Reflecting this concern, the census asked about the Mexican "race" in the 1930 Census.[10] For the most part, however, both the settled Mexican Americans and the circular migration of workers between Mexico and the United States were welcomed as economically necessary by Southwest ranchers and farmers and many local business people.

---

[9] The use of the 1890 Census figures was intended to be temporary until the Census Bureau could determine the national origins of the U.S. population. Anderson (2015, p.146) discusses the ensuing problems in detail.

[10] The addition of the Hispanic census category in the 1960s came from their own lobbying, although the category was an ethnic origin and not a racial group.

From a legal perspective, this changed in 1965 with the amendments to the Immigration and Nationality Act of 1952. These amendments were designed to remove the racist remnants of immigration law by eliminating restrictions on nationality or continent (Keely 1971). These legal changes could be seen as leveling the playing field for immigrants from any country, but one consequence of these changes was that immigration from the Western Hemisphere was no longer unrestricted. Although there were provisions for commuters across the Rio Grande bridges and for family reunification, the relatively free movement across the southern border was supposed to end. Mexican and Mexican-American families that had moved freely across the border to see one another were now expected to have documents and inspection. The new amendments took place in 1968.

By 1970 it was being alleged that a "flood" of illegal immigrants made it impossible to use demographic analysis to check for undercount in the 1970 Census. The reasoning was that the illegal immigrants were counted neither in birth records nor in migration statistics, because they were undocumented. This was true. What was also true was that a regular migration pattern had been redefined as illegal, and people who were formerly legally present in the country were also redefined as illegally present.

Besides people born in Mexico, there were many other people illegally present; nevertheless, Mexican illegal immigration became a familiar media meme. Others who were illegally present included visa overstayers from many countries, illegal entrants through Canada, and tourists who simply never left. The robust American economy was a magnet for workers of all kinds, skilled or unskilled.

The 1965 act also provided a preference for family reunification, so that many immigrants received visas to come to the United States to join family members. This provision was widely used. As a result of the 1965 act, the American immigration stream changed in its composition to include many more people from Asia—the Indian subcontinent, the Philippines, Korea, and many more countries. Immigration was opened to Africa and additional immigrants also came from the countries around the eastern Mediterranean, commonly called the Middle East.

As happened in the 1920s, this new immigrant stream came with different languages, cultures, and religions—Hinduism, Buddhism, but especially Islam. Many of these new immigrants were also dark-skinned. The same racial antagonisms that had been raised for centuries against African Americans were raised against some of the new immigrants. Many issues that had arisen in the 1920s were raised again, especially after the airplane hijackings, murders, and destruction of buildings on September 11, 2001. Many Americans lost their lives in what appeared to be a well-orchestrated attack by Muslim extremists, most of them from Saudi Arabia.

The new immigration from Asia and Africa meant many different languages were being spoken in America's largest cities. Heterogeneity of language strained local governments, especially as they sought translators for hospitals and municipal offices, appropriate services in schools, and culturally informed police and fire protection. To serve their constituents well, local agencies had to stretch already thin funding to provide for translators and sometimes for special training. School teachers certified in English as a Second Language found a soaring demand for their services.

As happened in the 1920s, it was alleged that crime followed these new immigrants wherever they resided, usually with the thinly veiled suggestion that the immigrants themselves were the actual criminals (Sampson 2008).[11] There is little evidence to support this charge. The argument was often heard that these "new" immigrants were costing far more than they brought to the country, even though most studies showed that the new immigrants paid more in taxes than the services they received (Karoly and Perez-Arce 2016).[12] The negative view of immigrants was reinforced, however, by the occasional and well-publicized cases of immigrants running crime rings, falsifying Social Security information, claiming welfare benefits to which they were not entitled, and smuggling drugs and other contraband to sell.

4.2   What About Citizenship?

The United Nations recommends citizenship as one of the questions to be asked in a census. The United States' stance on asking this question has varied over the years. In the 1950 Census, households were asked where each household member had been born. Question 14 read, "If foreign born, is the person naturalized?" This is not quite a citizenship question, because there are situations that could be covered by the combination of being born abroad but not naturalized. Someone born abroad to U.S. parents, for example, would have been born a citizen and not naturalized.

Since 1950 there has been no citizenship question asked of every household (U.S. Census Bureau n.d.).[13] In the 1960 Census, one in every four households completed the long form and were asked place of birth; birth country of the person's father; and birth country of the person's mother. Question 9 was "If foreign born, what is the person's mother tongue?" So, the question asked of the foreign-born became a language question and not a naturalization question. The additional questions allowed the analyst to determine the number of immigrants and the number of respondents who were the children of immigrants.

In 1970, once again only the households answering the long form were asked about place of birth. There were two versions of the long form, one for a 5% sample and one for a 15% sample, and questions asked in both samples could yield a 20% sample. Sample respondents were asked for place of birth, Hispanic origin, country of birth for the person's father and mother, and for persons born in a foreign country, whether they were naturalized and when they had come to the U.S. to stay. There was also a question about mother tongue. The naturalization question was

---

[11] Studies show that immigrant residential areas are actually safer than surrounding areas. In any event reported crime has decreased in the United States by 54% since 1993, according to FBI data (Gramlich 2019). The public, however, tends to believe that crime rates are increasing.

[12] Karoly and Perez-Arce suggest that state legislators see a positive benefit-to-cost ratio for immigrant-friendly policies such as in-state tuition. See also, Golash-Boza, T. (2009) in references.

[13] See, "Index of Questions" in references.

similar to the question asked in 1950, but it was asked only for a sample of households.[14]

A direct question about citizenship made its debut in the 1990 long form. Although the question was still asked of only a sample of the population, in 1990 and 2000 the question shifted from naturalization to citizenship. In 1990 and 2000, the question asked of a sample of households was "Is this person a citizen of the United States?" (The U.S. Census Bureau n.d.).[15]

Why did the Census Bureau stop asking everyone about citizenship after the 1950 Census? One procedural history of the census suggests that the reason was because census personnel believed that they could get adequate information from the alien registration forms that were required in those years before the 1965 Amendments.[16] The Immigration and Nationality Act of 1952 (the McCarran-Walter Act) required aliens to register once a year. The former Immigration and Naturalization Service (now ICE) also had records on immigrant entries and exits and also on the number of immigrants who were naturalized, and so the information was in principle available from administrative sources.

When undocumented immigrant flows increased in the 1970s and 1980s, the decision not to ask a citizenship question began to look more dubious. Without good data on the undocumented immigrant flow, politicians, public interest groups, and journalists of all stripes were free either to minimize the numbers who were illegally present or to exaggerate them beyond all belief. As with the undercount, the underlying issue was "if you didn't count them, then how can you know how many there are?"

4.2.1   Counting the Illegally Present Population

Demographers and statisticians both inside and outside the Census Bureau began to work on the problem of estimating the numbers of undocumented immigrants in the United States. The undercount was again a central issue. On the assumption that

---

[14] The same questions were asked of a sample of households in 1980, but questions were added about how well the respondent spoke English and what the person's ancestry was.

[15] See, "Index of Questions" in references. The 2010 Census did not have a long form because the American Community Survey replaced it.

[16] It is not clear how complete the alien registration was, but the Census Bureau argued that asking citizenship of the entire population "had become of less importance compared with other possible questions to be included in the census, particularly in view of the recent statutory requirement for annual alien registration which could provide the Immigration and Naturalization Service, the principal user of such data, with the information it needed." See, The U.S. Census Bureau (1983) "Censuses of Population and Housing: Procedural History 1940." Since the registration requirement was abolished, citizenship has been asked on various surveys and in some administrative forms. In the citizenship question lawsuit discussed below, the Department of Commerce informed the Court that the government believed it had citizenship data on 90% of the population, and that adding the question to the census could improve that coverage to perhaps 93% of the population.

many illegally present aliens would avoid as much contact with the government as possible, it seemed likely that the census would seriously undercount them.

There were some data sources that were helpful in making these estimates. Birth certificates are hard to avoid in the United States. One belief about the illegally present immigrants was that they would seek to have a child in the United States (so-called anchor babies) because anyone born in the United States is presumptively a citizen. When grown to adulthood, the baby born in the United States could sponsor family members under the family reunification provision of the immigration laws, so getting the birth certificate was precisely the contact with the government that the parents would want. Vital statistics bureaus in the states added a question about Hispanic origin, using wording similar to that in the census. This allowed demographic analysis of the undercount of Hispanic children. Of course, not all of their parents were illegally present; as we saw in the previous chapter, minority and low-income populations are most likely to be undercounted, and so an undercount of Hispanic children did not by itself point to the number illegally present.

Death certificates are similarly hard to avoid. The number of deaths can be classified by age, sex, and Hispanic origin. By dividing by age-sex-origin mortality rates, those deaths can be expanded to the size of the population from which they must have come. This was another approach to estimate the size of the Hispanic population, but it ran into two issues. The first was that with deaths, as with births, being Hispanic was by no means synonymous with being illegally present. The second issue was more technical: which death rates should be used? While mortality by age and sex shows grossly similar patterns for all groups in the United States, there are enough differences in the patterns to make the resulting estimates questionable.

In 2018, the Office of Immigration Statistics estimated that nearly 6.6 million people from Mexico were in the United States illegally as of 2015 (Baker 2018). But there were illegal aliens from other countries as well, and even fewer ways to estimate, say, the number of illegally present Iranians, or Chinese, or Indians. Other methods were tried, especially comparisons of survey data. Most of these methods were still subject to the same objections that people who were illegally present would seek to avoid the survey or perhaps to give an incorrect answer (Sullivan et al. 1984).

4.3   A Sharp Turn in Immigration Policy

The 2016 presidential campaign signaled that immigration would be a marquee issue for the Trump Administration. The signature program was building a wall between the United States and Mexico, and initially the anti-immigrant theme appeared to target Mexicans. Not long after Donald Trump was inaugurated, a number of executive orders were issued that affected which immigrants would be allowed into the country. Many of these orders were challenged in court, especially those which seemed intended to prevent Muslims from entering the United States by targeting certain Muslim-dominant countries. What became clear, however, was that a general reduction in immigration, both legal and illegal, was intended.

Visas became more difficult to get, even for international students legitimately admitted to American universities to study or for workers legitimately hired by American employers. This visa action did not require any new laws; it just required instructions to the consular offices to be more probing and less ready to approve visas.

Asylum seekers were turned away from the southern border and asylum seekers from other parts of the world were generally not accepted. The Syrian refugee crisis, which resulted from a civil war within Syria, was met with at best tepid response from the United States. Economic refugees fared no better. Several "caravans" of asylum-seekers from Central America were turned back to Mexico. These would-be refugees were fleeing crime and chronic poverty, and many were in legitimate fear for their lives if they returned home. In a particularly well-publicized move, children of asylum seekers were separated from their parents and both groups were placed into detention camps that looked a great deal like jails. There were deaths of both children and adults while being detained.

Another set of immigration issues involved children who had grown up in the United States following their parents' illegal residence. DACA stands for Deferred Adjudication for Childhood Arrivals. It was a program created by President Barack Obama's executive order in 2012. The DACA applicants were foreign-born but had been brought to the United States as children, when they were too young to have any say concerning their entry into the United States. The parents who brought them were illegal entrants or became illegal later, typically through visa overstaying.

The DACA cases aroused a great deal of sympathy and concern, and a case concerning DACA has reached the U.S. Supreme Court. DACA participants, sometimes called Dreamers for seeking the American Dream, are often depicted in the media as being from Mexico or another Spanish-speaking country. In fact, their parents have come from many different countries.[17]

DACA applicants were given permission to remain and to work in the United States under a set of strict provisions. They were required to apply for the status, which involved giving their names and location information to the government. They were finger-printed, underwent criminal background checks, and were expected to make progress in their educations or to be at work. Any infraction— including a DUI, in some cases—could lead to their deportation. In most cases, their family members were protected from deportation while the DACA applicant remained eligible. Journalistic profiles of DACA recipients were usually appealing, and the program was viewed positively in most quarters.

DACA participants might be viewed as documented, in the sense that the government knows where they are and has given them documents granting them a tempo-

---

[17] The author has personal experience with DACA university students, only one of whom was from a Spanish-speaking country. The others were from a variety of countries in Asia, Africa, and Europe, and most had parents who were defined as visa overstayers. In addition, many of the parents had believed themselves on the path to citizenship, but through various bureaucratic fumbles at ICE, their records applying for green cards were lost. In some countries of origin, it is difficult if not impossible to replace the documents that had been lost, so the parents were left in an untenable situation and so were their children.

rary permission to stay. The Trump Administration has vacillated on the future of the DACA program, with the President sometimes expressing sympathy for the DACA participants while his administration seeks agency or judicial action to remove them. Some DACA participants have been deported, and others live in a constant state of worry about whether they will be deported. Meanwhile, both Congress and the administration have been stalemated on a long-term solution.

All of these were immigration issues: building the Wall, tightening the numbers of visas, banning certain countries, turning back asylum seekers, deporting DACA participants. Each of them received saturation-level coverage in print and electronic media. Immigrant communities carefully read and watched these stories, and apprehensions about the United States and its government grew. The fear was felt by legally present as well as illegally present individuals.

In earlier censuses, the U.S. government has reassured immigrants, even those illegally present, that participating in the census would involve no negative consequences for them. In 2010, there was even a publicly declared moratorium on immigration raids and similar activities. The moratorium was explicitly advertised as a means of reassuring immigrants about their participation in the census. These efforts indicated that census and homeland security officials were aware that participating with the federal government aroused concerns among immigrants.

This background of concern and even fear became an overriding issue with the proposed citizenship question.

4.4   The Issue of a Citizenship Question On the 2020 Census

On March 26, 2018, Secretary of Commerce Wilbur L. Ross, Jr., announced that a question on citizenship would be added to the 2020 Census. This date was well after the various content tests of the census had begun. It was not the case that the question itself had never been pre-tested. As mentioned above, a citizenship question has been asked in the American Community Survey for more than a decade.

The conditions of taking the survey are different from the census, however. The census uses self-response as much as possible. The more follow-up required by census workers to pursue non-response, the more expensive the census becomes. Given cuts in census funding, there was real concern that anything that potentially increased non-response would also increase costs. In a survey, an interviewer makes the contact and asks the questions. A reluctant respondent can be reassured, and the interviewer can gently probe to get more complete answers. While respondents can refuse to answer, the ACS, like the census, requires answers by law.

Given the long running interest in reducing the undercount (see Chap. 3), the news that relatively late in preparation there was to be an eighth question added to the census created a variety of concerns: instrumental costs such as developing and printing new questionnaires and, most importantly, the mission-critical effect the citizenship question could have on the undercount.

In most census cycles, the addition of a question late in the census development process would have been sufficient to cause concern among the professional staff. In 2018, given the background of the administration's animus toward immigration, the potential addition of a census citizenship question raised fears among many concerned groups about a larger undercount (if non-citizens refused to participate) and about the privacy of the census (if non-citizens did participate). The public discussion of the effort to add a citizenship question to the 2020 Census intensified public mistrust (Levitt 2019).

In the absence of citizenship data from the various census tests, demographers and statisticians both inside and outside the Census Bureau began to ransack other datasets to make estimates of the effects of a citizenship question on the census. Several analyses and experiments suggested that the question would lead to an increase in undercount and data inaccuracy.

Matthew Baum and associates (Baum et al. 2019), researchers from Harvard's Shorenstein Center for Media, Politics, and Public Policy, designed a large survey experiment with a questionnaire that resembled the 2020 Census in appearance and content. Half of the sample, approximately 4500 people, received the approved census questions and the other half received a questionnaire with the citizenship question added. Based on comparing the responses from the two conditions, they concluded that the citizenship question could lead to potential underreporting of between 5.7 million and 6.2 million Hispanics because of the significant increase in questions skipped. The effect among Hispanic respondents was especially strong.

J. David Brown and his associates (Brown et al. 2018, 2019) are Census Bureau researchers who concluded something similar using an official source, the American Community Survey. The ACS contains a citizenship question. From administrative records (not the ACS, but other government records) they knew which ACS respondents were recorded as citizens. They also knew which respondents were recorded as non-citizens. They found that 2017 ACS respondents who had been recorded as non-citizens were more likely to skip the citizenship question or to answer (incorrectly) that they were citizens. The researchers also compared the mail response rates for households that had returned the 2010 census form (with no citizenship question) with the same households that had fallen into the 2010 ACS sample (with a citizenship question). Based on their comparison of the two response rates for the very same households, they concluded that the addition of a citizenship question could cause an 8% drop in response rate for households containing a non-citizen. This drop would extrapolate to a drop of 2.2% in the overall response rate for the census. The results would be an increase in follow-up costs and very likely a drop in the quality of the census.

William P. O'Hare (2018), one of the foremost experts on undercount, also studied the responses to the ACS citizenship question. He found that the citizenship question has the highest non-response among the items on the ACS. In 2016, the citizenship question was skipped by 6% of respondents versus a high of 1.2% for any other question that is asked both on the ACS and in the census. The non-response rate for the citizenship item increased over the decade, and it was especially sensitive for racial and ethnic minorities, immigrants, and residents of central cities.

The non-response was higher for self-response questionnaires, especially internet response, and the non-response was much higher in some cities and states. He concludes, as did the Census Bureau researchers, that the non-response could increase census costs or decrease quality or both.

One study went the other way. Victoria A. Velkoff (2019) is Associate Director for Demographic Programs at the Census Bureau. She reported on a field experiment of 480,000 households, half of which were randomly assigned to receive the citizenship question along with other census questions, while the other half did not receive the citizenship question. The contact procedure was similar to the one that will be used in 2020, including the availability of telephone help. There was a smaller mail self-response rate for those who received the citizenship question (16.0 percent versus 15.7%) and no difference for internet and telephone responses. There was a statistically significant difference (0.3%) between the two forms in the proportion of respondents who identified as Hispanic. Nevertheless, Velkoff concluded that the addition of the citizenship question would not have increased non-response follow-up staffing costs. Notably, no non-response follow-up was pursued in this experiment.

A number of lawsuits were filed to prevent the citizenship question from being added to the census, and the issue reached the Supreme Court during its 2018–2019 term (Department of Commerce v. New York 2019). The 92-page decision of the Supreme Court ultimately turned on the disputed issue of the Secretary's true reasons for adding the question late in the process of census preparation. The trial record and parts of the Justices' four separate opinions repeatedly alluded to the statistical community's concerns about whether the citizenship question would be answered accurately or answered at all.

The Secretary of Commerce has been delegated authority for the conduct of the census, including the questions to be asked. Congress requires the Secretary to submit a list of the questions to be asked no later than 2 years before the census date (U.S. Census Bureau 2017).[18] This requirement exists because there is often some controversy in Congress about how many questions are being asked and whether there is too much burden on the people, and also whether the questions are necessary because the answers could be collected in some other way. For example, asking questions about education and occupation in the American Community Survey is a lower burden on the public than asking those same questions of every respondent in the census would be. The Census Bureau usually justifies each question by pointing to its required use by a statute or by an agency.

In the case of the citizenship question, the Secretary stated that the Department of Justice required the citizenship question for purposes of enforcing the Voting Rights Act. This reason was surprising because the Attorney General had not so far seemed very interested in enforcing the Voting Rights Act.

---

[18] See, "Subjects Planned for the 2020 Census and the American Community Survey: Federal Legislative and Program Uses" in references.

Through the process of discovery in the early stages of the litigation, the plaintiffs learned that it was the Secretary of Commerce who had in fact asked the Department of Justice to ask for the citizenship data. The Department of Justice eventually did so, and then the Secretary of Commerce called for the question to be asked.

The Supreme Court's decision ultimately turned on the issue of the Secretary of Commerce's reason for asking the citizenship question. A majority of the Court agreed that the Secretary's stated reason had been a pretext, and therefore the question was not added to the 2020 Census. It seemed clear that there would be no barrier to adding the question to the 2030 Census, and that there would have been no barrier in 2020 but for the pretextual reason that had been given.

A pretext suggests that there was a real reason for asking the question, but a hidden, unstated reason. The Court did not have to tell the public what the real reason was, and perhaps the Court did not know. But there were many pundits who were happy to fill in the blanks.

One line of reasoning went like this: a citizenship question would presumably deter Hispanics and immigrants from answering the census, or at least from answering the citizenship question. It would especially deter anyone illegally present in the country. Thus it would increase the undercount, with all the consequences that were described in the preceding chapter. By this reasoning, the Secretary of Commerce was willing to trade a higher undercount for an improved specification of citizens. This reasoning flies in the face of decades of work by the Census Bureau to reduce the undercount. It seems ingenuous of the Secretary to believe that this trade-off was in line with the twin goals of a more complete and accurate census.

A second line of reasoning, however, soon emerged. Thomas Hofeller, a recently deceased Republican strategist from North Carolina, left behind some 75,000 files on four hard drives and 18 thumb drives. These files were discovered by his daughter. These papers had apparently previously been shared with members of the Republican party. His daughter shared them with the advocacy group Common Cause.

Mr. Hofeller's work dating from 2015 supported the addition of a citizenship question to the 2020 Census (Wines 2019).[19] He had hypothesized that citizenship data could be used in the redistricting process to dilute the political power of Democrats. The mechanism was that state legislatures should redistrict based on citizenship and not just population counts. It is settled that apportionment among the states is based on the total population count, but state legislatures have some leeway in how to draw districts within the state (National Conference of State Legislatures 2019).[20]

If redistricting is based on total population counts, then clusters of non-voters can create extra districts for the cities and counties in which they reside at the expense of the localities where only voters reside. These non-voter clusters could be

---

[19] Evidently Mr. Hofeller also wrote the key portion of a draft Justice Department letter that claimed the Justice Department needed citizenship data to enforce the Voting Rights Act.

[20] State legislatures must comply with redistricting requirements from the Supreme Court's interpretation of the U.S. Constitution and from the Voting Rights Act. Beyond those requirements states are allowed to adopt their own redistricting criteria.

made up of any group that does not or cannot vote, such as neighborhoods with many children, a district containing a large prison, or a district with many immigrants who are not or cannot be naturalized. By redistricting the state using total population counts, the legislature amplifies the vote of the voter whose district has many people who cannot vote, such as immigrants or children.

In central city areas, where many immigrants live, the voting population often belongs to minority groups and has low income. Some minority populations also have higher fertility rates than white populations. Based on voting data, areas with these population characteristics tend to vote Democratic. To Mr. Hofeller, using total population counts seemed to be amplifying the vote of Democrats.

If instead the legislature were to redistrict based only upon the counts of voting-age citizens, then the districts might look quite different. Older, white, suburban voters, who are statistically more likely to vote Republican, would have their votes amplified because the non-citizens would not be included in the redistricting computations.

This second line of reasoning suggested that partisan interests were the real reason for the effort to add the citizenship question. Redistricting based upon citizenship might appeal to Republican-dominated legislatures, but those legislatures could not act upon that appeal because they did not have information about how many citizens they had and where those citizens lived. Adding a citizenship question to the census would provide the means to sort the enumeration by voting-age citizens versus all other residents for every level of census geography. These geographic levels could then be combined to form legislative districts.[21]

4.4.1 Enter Gerrymandering

Gerrymandering refers to drawing districts to further the interests of one party, and it carries the connotation of being unfair. The term comes from Massachusetts Governor Elbridge Gerry, who after 1812 redrew Massachusetts districts to favor the Democratic-Republican party and to disadvantage the Federalists. To cartoonist Elkanah Tisdale the shape of the districts resembled the parts of a salamander; thus, it was called a "gerrymander."

In Gerry's day, redrawing districts by hand was time-consuming and laborious. Today with computers an analyst can set parameters for the districts and have multiple options ready in seconds. The basic idea behind gerrymandering is to give one's own political party the advantage and disadvantage the opposing party.

---

[21] Mr. Hofeller was working with Republicans, but whatever party is in control of the legislature can use such data to their advantage. To prevent such manipulation, some states have recently moved to an independent or bipartisan redistricting commission or they are considering doing so. States currently using a redistricting commission are Alaska, Arizona, California, Colorado, Hawaii, Idaho, Iowa, IdahoIowa, Michigan, Missouri, Montana, New Jersey, New York, Ohio, Pennsylvania, Utah, Vermont, and Washington (Associated Press 2019).

There are two basic approaches to doing gerrymandering. One is packing, which means crowding all of the members of the opposing party into as few districts as possible. The other is cracking, which is fragmenting the numerical strength of the opposing party by drawing lines through their residential areas of strength, then appending those areas to different districts that contain larger numbers of voters from one's own party.

Of course, the speculation about who will vote which way has to be based upon historic voting and residential patterns. A so-called wave election, in which one party performs dramatically better than it previously had performed, may throw off the calculations of those who drew the maps.

4.4.2   Population Shifts and Gerrymandering

In 1790 the United States was predominantly a rural country huddled for the most part on the shores of the Atlantic. By 1970 the United States was predominantly urban and spread across the continent, with Alaska, Hawaii, and the five island jurisdictions included. The number of states had increased from 13 to 50. And the American population was restless, with a steady movement from east to west and then from northeast to southwest.

In general, the population growth of the Southwestern states has increased their national political power. And although the entire country had grown from 179 million in 1960 to 309 million in 2010, the growth has by no means been even. After the 1960 census, the states of Arizona, California, New Mexico, Nevada, and Texas had 67 members of Congress.[22] Fifty years later, following the 2010 Census, they had 105 members. As Table 4.1 shows, the percentage of the House of Representatives from these states grew from 15.4% in 1960 to 24.1% in 2010. It is projected to increase to 25% after the 2020 Census.

The advent of widespread air-conditioning, the relocation of some manufacturing, the growth of trade with Mexico and the rest of Latin America, the development of high-tech hubs in the Southwest, and the attraction of Southwestern cities for young and old alike have aided the rapid growth. The oil and ranch economy of earlier years is still there, but now joined by many other industries. Migration to the Southwest was not only from Mexico; the Southwest has had positive net migration from other regions in the United States. In addition, differential fertility has favored the Southwest. But the most politically charged difference has been greatly increased international migration, legal and illegal, to the Southwest. Given the additional districts that would need to be drawn as the Southwest expanded, legislatures there had an unusual opportunity to experiment with redrawing district lines with minimal disadvantage to incumbents.

---

[22] "Southwest" is not a census designation. I chose these states because they were border states or near-border (i.e., Nevada) states that generally experienced the growth described in the text.

Table 4.1 This table shows how the distribution of seats in the U.S. House of Representatives has changed since 1960 in what this chapter describes as "The Southwest" states of Arizona, California, Nevada, New Mexico, and Texas. For data sources see Mills (2001), Burnett (2011), and Cea (2018)

	Seats	Percent of House	Seats	Percent of House	Seats	Percent of House	Seats	Percent of House	Percent increase 1960–2020
Year	1960	1960	2000	2000	2010	2010	2020	2020	
Arizona	3	0.7%	8	1.8%	9	2.1%	10	2.3%	200%
California	38	8.7%	53	12.2%	53	12.2%	52/53	12.2%	39%
Nevada	1	0.2%	3	0.7%	4	0.9%	4	0.9%	300%
New Mexico	2	0.5%	3	0.7%	3	0.7%	3	0.7%	50%
Texas	23	5.3%	32	7.4%	36	8.3%	39	9.0%	56%
Number of House members from The southwest	67	15.4%	99	22.8%	105	24.1%	108/109	24.8%/25%	63%

Mr. Hofeller, the Republican elections strategist, left files that revealed an "exhaustive" analysis of Texas state legislative districts (Wines 2019). Drawing the maps using voting-age citizen populations "would be advantageous to Republicans and non-Hispanic whites." The maps would also dilute the political power of Hispanics. After children and non-citizens were removed from the calculations, the traditionally Democratic districts would be forced to expand in area to create a district numerically equivalent to the Republican district. In effect, the one person, one vote requirement would become one adult citizen, one vote.

4.4.3 One More Supreme Court Case

The day before ruling on the census citizenship question, the Supreme Court issued a decision on gerrymandering (*Rucho v. Common Cause* 2019). In a nutshell the gerrymandering decision turned on the issue of whether legislatures could develop voting districts in such a way as to maintain a majority of voters from the dominant party and therefore maintain their power in the state legislature for a decade. The Supreme Court in a split decision denied that the courts had a role to play in a purely political decision such as redistricting (*Rucho v. Common Cause* 2019).

The Court's opinion was that the political aspects of redistricting were reserved to the elected branches of government. Pure racial gerrymandering was not allowed by this decision because it was forbidden by the Voting Rights Act. The Fourteenth Amendment provided the constitutional basis for the earlier requirements for districts to have numerical equality, compactness, and contiguity. But beyond those constraints, state legislatures had broad discretion to be partisan in redistricting.

Rucho is an important legal development in juxtaposition with the citizenship question decision, which was issued the next day. In the census case the Court held that the Secretary of Commerce had violated the Administrative Procedure Act by offering a pretextual reason for adding the citizenship question, namely enforcing the Voting Rights Act. If the Secretary had been more forthcoming with his real reasoning, the opinion suggested, then there would be no judicial impediment to including the citizenship question.

If legislatures could redistrict using number of *citizens* instead of number of *inhabitants*, then districts could be more precisely drawn to benefit one party (Henderson 2019; Edmonston and Wines 2019). To do this, however, the state legislatures need data on citizens, not inhabitants. In short, if the Secretary had given this reason, the Court's decision on gerrymandering—that the Court had no role to play in a purely political process—would imply that its decision on the citizenship question would have been different, and the citizenship question would have been included on the census forms in 2020. Arguably even a nakedly partisan reason would be sufficient if the other requirements of the Administrative Procedure Act were met.

If voting-age citizenship became the basis for redistricting within states, what would change? Census counts of all inhabitants would still be the basis for reapportioning Congress and for distributing the funding for federal programs. But within the states, districts could look very different. As we shall see in Chap. 6, this latter result could still occur after the 2020 Census, even without the citizenship question.

4.5 Summary

In the preparation for the 2020 Census, two reasonable objectives were pitted against each other: reducing the undercount and providing a count of citizens. The tension between the two stems from recent changes in the policies and practices of federal executive agencies that are perceived as anti-immigrant. These actions have been widely publicized. Both observers and the immigrants themselves have expressed fear that seeming hostility to immigration will increase the undercount. There is at least some evidence that adding a citizenship question would increase the undercount among Hispanics and among immigrants.

Two Supreme Court cases have sharpened the debate on these topics for the 2030 Census. The Court did not permit the addition of a citizenship question for all households in the 2020 Census, but it did rule that purely partisan gerrymandering was not an appropriate topic for judicial intervention. Thus, the stage is set for a future redistricting of state legislatures that is based only on voting-age citizens and eliminates the other inhabitants who are counted in the census.

References

Anderson M (2015) The American Census: A Social History, 2nd edn. Yale University Press, New Haven, CT. 204, 218, 253, 255, 259–260

Baker B (2018) Population estimates: illegal alien population residing in the United States: January 2015. Department of Homeland Security, Office of Immigration Statistics. https://www.dhs.gov/sites/default/files/publications/18_1214_PLCY_pops-est-report.pdf. Accessed 20 Dec 2019

Baum MA, Dietrich BJ, Goldstein R, Sen M (2019) Estimating the effect of asking about citizenship on the U.S. Census: results from a randomized controlled trial. Harvard University Kennedy School, Shorenstein Center on Media, Politics, and Public Policy. Retrieved from https://shorensteincenter.org/estimating-effect-asking-citizenship-u-s-census/. Accessed 21 Dec 2019

Brown JD, Heggeness ML, Dorinski SM, Warren L, Yi M (2018) Understanding the quality of alternative citizenship data sources for the 2020 census (Center for Economic Studies Working Paper Series No. 18-38R). Washington, DC: U.S. Census Bureau. https://doi.org/10.5281/zenodo.3298987. Accessed 4 Dec 2019

Brown JD, Heggeness ML, Dorinski SM, Warren L, Yi M (2019) Predicting the effect of adding a citizenship question to the 2020 census. Demography 56(4):1173–1194. https://doi.org/10.1007/s13524-019-00803-4

Burnett KD (2011) Table 1: apportionment population based on the 2010 census and apportionment of the U.S. House of Representatives: 1910 to 2010. Congressional Apportionment: 2010 Census Briefs. https://www.census.gov/prod/cen2010/briefs/c2010br-08.pdf. Accessed 28 Oct 2019

Cafferty PSJ, Chiswick BR, Greeley AM, Sullivan TA (1983) The dilemma of American immigration: beyond the golden door. Transaction Books, New Brunswick, NJ

Cea B (2018) Census Projections for 2020 Congressional Reapportionment. Potential Shifts in Political Power after the 2020 Census. Brennan Center for Justice. https://www.brennancenter.org/our-work/research-reports/potential-shifts-political-power-after-2020-census. Accessed 28 Oct 2019

Department of Commerce v. New York No. 18–966 (2019). https://www.supremecourt.gov/opinions/18pdf/18-966_bq7c.pdf. Accessed 4 Dec 2019

Edmonston C, Wines M (2019) Official's testimony adds to rancor around census citizenship question. New York Times June 25. https://www.nytimes.com/2019/06/25/us/politics/census-citizenship-question.html. Accessed 18 Dec 2019

Golash-Boza T (2009) The immigration industrial complex: why we enforce immigration policies destined to fail. Sociol Compass 3(2):295–309

Gramlich, J. (2019). 5 facts about Crime in the U.S. Pew Research Center, October 17. https://www.pewresearch.org/fact-tank/2019/10/17/facts-about-crime-in-the-u-s/. Accessed 3 Dec 2019

Henderson T (2019) How the citizenship question could reshape state politics. Pew Stateline June 28. https://www.pewtrusts.org/en/research-and-analysis/blogs/stateline/2019/06/28/how-the-citizenship-question-could-reshape-state-politics. Accessed 20 Dec 2019

Karoly LA, Perez-Arce F (2016) Understanding the costs and benefits of state-level immigration policies. RAND Research Brief (March). https://www.rand.org/pubs/research_briefs/RB9923.html. Accessed 4 Dec 2019

Keely CB (1971) Effects of the immigration act of 1965 on selected population characteristics of immigrants to the United States. Demography 12:179–191

Levitt J (2019) Nonsensus: pretext and the decennial enumeration. 3 ACS Sup. Ct. Rev. 59. https://ssrn.com/abstract=3469935. Accessed 4 Dec 2019

Mills KM (2001) Table 1: apportionment population based on census 2000 and apportionment of U.S. House of Representatives 1900–2000. Congressional Apportionment: Census 2000 Brief. https://www.census.gov/population/apportionment/files/2000%20Apportionment%20Brief.pdf. Accessed 28 Oct 2019

National Conference for State Legislatures (2019) Redistricting criteria. http://www.ncsl.org/research/redistricting/redistricting-criteria.aspx. Accessed 16 Dec 2019

O'Hare WP (2018) Citizenship question nonresponse: Demographic profile of people who do not answer the American Community Survey citizenship question. Washington, DC: Georgetown Center on Poverty and Inequality. http://www.georgetownpoverty.org/wp-content/uploads/2018/09/GCPI-ESOI-Demographic-Profile-of-People-Who-Do-Not-Respond-to-the-Citizenship-Question-20180906-Accessible-Version-Without-Appendix.pdf. Accessed 6 Dec 2019

Prewitt K (2013) What is your race? The census and our flawed efforts to classify Americans. Princeton University Press, Princeton, NJ

Rucho v. Common Cause, 139 S. Ct. 2484 (2019)

Sampson RJ (2008) Rethinking crime and immigration. Contexts 7(1):28–33

Sullivan TA, Gillespie FP, Hout M, Rogers RG (1984) Alternative estimates of Mexican-American mortality in Texas, 1980. Soc Sci Q 62(June, 1984):609–617

The Associated Press (2019) Number of states using redistricting commissions growing. https://apnews.com/4d2e2aea7e224549af61699e51c955dd. Accessed 22 Dec 2019

The U.S. Census Bureau (1983) Censuses of Population and Housing: Procedural History 1940. https://www.census.gov/history/pdf/1940proceduralhistory-12617.pdf. Accessed 20 Dec 2019

The U.S. Census Bureau (2017) Subjects planned for the 2020 census and the American Community Survey: Federal Legislative and Program Uses. https://www2.census.gov/library/publications/decennial/2020/operations/planned-subjects-2020-acs.pdf. Accessed 16 Dec 2019

The U.S. Census Bureau (n.d.) Index of questions. https://www.census.gov/history/www/through_the_decades/index_of_questions/1950_population.html. Accessed 6 Dec 2019

Velkoff VA (2019) 2019 Census test preliminary results. https://www.census.gov/newsroom/blogs/random-samplings/2019/10/2019_census_testpre.html?utm_campaign=&utm_medium=email&utm_source=govdelivery. Accessed 1 Dec 2019

Wines M (2019) Deceased G.O.P. Strategist's Hard Drives Reveal New Details on the Census Citizenship Question. New York Times (May 30). https://www.nytimes.com/2019/05/30/us/census-citizenship-question-hofeller.html. Accessed 16 Dec 2019

Zolberg A (2006) A nation by design: immigration policy in the fashioning of America. Russell Sage Foundation/Harvard University Press, New York

Chapter 5
Privacy and Data Protection

Abstract Since 1850 there has been increased attention to data confidentiality in the census. Title 13 of the U.S. Code makes confidentiality mandatory. Census personnel take lifelong oaths to keep the data confidential, and many statistical protections have been developed to mask data about individuals. The digital era has made data confidentiality more difficult to assure, in part because the abundance of commercially and publicly available data that is not protected can potentially be used to re-identify a respondent within the census. An algorithmic process called differential privacy will be deployed for the 2020 Census. By using "noise injection," this process increases the uncertainty of identifying a particular person. The scientific community has raised the issues of whether the increased privacy is an acceptable trade-off for accuracy, and of what effects a differentially private census might have on other data bases that depend upon the census. The Census Bureau is also preparing for cyberthreats including hacking, fake websites, misinformation, and disinformation.

Keywords Confidentiality · Privacy · Differential privacy · Cyberthreat

It was time to purge the hacker from the U.S. government's computers. After secretly monitoring the hacker's online movements for months, officials worried he was getting too close to critical information and devised a plan, dubbed "the Big Bang," to expel him. Trouble was, with all their attention focused in that case, they missed the other hacker entirely (Tucker 2016).[1]

The federal government faces enormous issues with cybersecurity (Wolff 2019).[2] Every day malicious actors launch innumerable attacks against secured and unsecured government databases. In the attack referenced above, a foreign power successfully attacked the Office of Personnel Management (OPM) and swiped millions of personnel records, fingerprints, and identifying data from federal employees and other citizens who had security clearances. The federal government is currently paying for 3 years of credit monitoring for the victims, and there is still concern for how their data might be misused down the road (Naylor 2016).

Successful data breaches are so common now that they are barely newsworthy. Their very ubiquity has made members of the public suspicious about putting personal information anywhere that cyberthieves could steal it. The news that the 2020 Census would move toward internet response as the preferred form of response has concerned people who care in general about their information being unsafe. And then there are people who care in particular that the government could misuse the data. As it turns out, the Census Bureau itself has a concern related to reverse engineering of census data to reveal individual information. Each of these issues will be addressed in this chapter.

5.1 The Terms Privacy and Confidentiality

As used in conjunction with census data, the terms privacy and confidentiality are closely related (Singer et al. 1993). Confidentiality may be considered a practice by which respondents' names and other identifying pieces of information are not associated with specific individuals.[3] Singer and her colleagues report on some early field experiments that suggest the belief in confidentiality has minor effects in encouraging survey response rates and quality; on the other hand, they also note that

---

[1] Tucker, E. (2016). "OPM hack: Congressional report spells out missed chances to stop cyber breach." *Seattle Times*. September 7. https://www.seattletimes.com/business/missed-opportunities-to-stop-opm-cyber-breach-spelled-out/

[2] Given the cybersecurity threats in many industries, the federal government is beginning to have difficulty in retaining its top cybersecurity experts.

[3] Census data are tabulated and published, and so an individual's responses become public, but only as they are blended into statistical summaries with the responses of thousands or millions of others. So, a thirty-year-old woman living in Austin, Texas, might be listed in a table on the population of Austin classified by age and sex, but there would be enough people in that one cell of the table (women aged 30–34) that no single individual could be identified.

elaborate assurances of confidentiality may backfire, increasing the suspicion of the potential respondents and reducing their readiness to respond.[4]

Confidentiality also implies a commitment to keep access to the data limited to individuals with a right to know. Sharing beyond that group violates the expectation of confidentiality. As we shall see, over time census professionals have operated under increasing confidentiality protections, including legal penalties for violating confidentiality.

Privacy refers to information that a person wishes to keep secret. Violating the sphere of privacy is perceived as intrusive, and many individuals find intrusive questioning by the government to be particularly offensive. Early concerns about privacy are reflected in the Fourth Amendment of the Constitution, which provides for protecting one's home and possessions from unreasonable searches and seizures.

People differ among themselves in the things that they think no one (perhaps especially the government) has a right to know. Some people are sensitive about their age or their racial identity. From time to time privacy concerns have been raised about questions in the census of housing, such as how many bathrooms a home has. Many Americans feel very private about any financial information, such as their annual income or their level of indebtedness.

There are recent indications that Americans have a heightened interest in privacy. This increase in privacy concerns has paralleled the ubiquity of electronic technology, and the concern has moved beyond government to concern about other sources such as technology companies, nosy neighbors, and foreign entities unobtrusively observing or listening. These concerns are expressed in terms of resistance to intensified security screenings and surveillance technology. The revelations that internet providers and some electronic home gadgets can "snoop" on families and their behavior have been unsettling. The concern with privacy seems even greater in some European countries. The European Union has very strict privacy regulations, and practices such as inserting cookies on one's computer for visiting a website are regulated much more strictly there than in the United States.

How much do these concerns affect census response? In the 1990 Census, Americans who had a low level of confidence in Census Bureau confidentiality were more than twelve percentage points less likely to mail back their census returns than were Americans who had a high level of confidence that their information would not be shared (Fay et al. 1991). This was especially noticeable in white respondents. Black respondents were more likely than white respondents to return the form even if they reported higher levels of mistrust (Singer et al. 1993, p. 479).

A difference of twelve percentage points in response rates translates into millions of dollars in costs for non-response follow-up. Given the great explosion of internet availability, wi-fi, and electronic devices of all sorts since the 1990 Census, confidentiality and privacy are important concerns for the 2020 Census.

---

[4]The Census Bureau's own experiment with confidentiality-related messages in the cover letter for the initial census mailing reveals no statistical differences between the experimental and control (2010 Census statement) language. See Hill et al. (2012).

5.2   A History of Data Confidentiality

The first census in 1790 had no confidentiality; in fact, it was about as public as it could be and for good reason. With most of the population rural and many of them poorly housed, the U.S. marshals who took the census could not be certain that they had located every household. The marshals posted the results in town squares and other public places so that individuals could assure themselves that they had been counted. Thomas Jefferson, who as Secretary of State had the oversight of the first census, is enumerated in Albemarle County, Virginia, somewhat strangely. His name is scrawled in the margin of the enumeration sheet. It appears that his name was an afterthought, but another interpretation is that he was missed and took the opportunity to add his name to the posted return in the public square (Sullivan 2020).[5]

Through the 1800s, there were early indications of confidentiality concerns. Besides enumerating households, the Census Bureau and its forerunner agencies also conducted censuses of economic activity. The return rates for the 1840 Census of Manufacturers were so low that the U.S. marshals were instructed to tell the respondents that their answers would be confidential (U.S. Census Bureau 2019b).[6] It seems likely that the businesses being enumerated were concerned about competitors and the information their competitors could gain about them from the census. Even without knowing trade secrets, a competitor could be advantaged by information such as number of workers, receipts, and so on.

By the 1850 Census, the marshals were instructed to consider all census responses to be confidential. After 1870, the marshals no longer conducted the census. Using trained enumerators instead of law enforcement workers might have helped reassure the populace to cooperate. By the 1880s there were stiff fines for census workers who violated their oaths of secrecy. Meanwhile, however, census results were for sale and state and municipal governments could receive lists of names and details from the census. The Permanent Census Act of 1902 authorized the Census Director to furnish "copies of so much of the files or records of the eleven decennial enumerations from 1790 to 1890, inclusive, as may be requested" by a governor or mayor.

---

[5] Sullivan (2020, note 3). writes: "Compare [Jefferson's] known signatures at https://images.search. yahoo.com/search/images;_ylt=AwrJ7FeeL45d8BôA.1ZXNyoA;_ylu=X3oDMTEyMmg2cmYw BGNvbG8DYmYxBHBvcwMyBHZ0aWQDQUJBQ0tfMQRzZWMDc2M-?p=thomas+jefferso n+signature&fr=tightropetb with the image of the enumeration page from the 1790 Census, https:// www.archives.gov/research/census/presidents/jefferson.html, which appears consistent with [Jefferson's] other signatures but inconsistent with the hand in which the other entries are written. Returns from the 1790 Census had to be publicly posted for correction, and one possibility is that ... Jefferson was merely adding himself to the count."

[6] See, "The 2020 Census and Confidentiality" in references. Much of the information in the following paragraphs is taken from this source. Another useful source is U.S. Census Bureau (2019a) "Events in the Chronological Development of Privacy and Confidentiality at the U.S. Census Bureau" in references.

A major new step in publicizing the commitment to confidentiality came in March 1910, when there was a Presidential Proclamation by President William Howard Taft saying "The sole purpose of the Census is to secure general statistical information. ...every employee of the Census Bureau is prohibited, under heavy penalty, from disclosing information which may thus come to his knowledge..."

Wartime, however, brought different exigencies, and the Census Bureau was legally permitted by wartime laws to share quite a lot of information during World War I. In 1916 the Navy received from the Census Bureau the names and addresses of 30,000 large manufacturing plants, and in 1918 and 1919 the Department of Justice and local registration boards received the ages of men who failed to register under the selective service law. That same year the Census Bureau provided mailing lists for Liberty Loans and it gave the U.S. Fuel Commission a list of cotton gins in Georgia (U.S. Census Bureau n.d.).[7]

The most contested revelations of census data, however, came during World War II. Under the Second War Powers Act (1942), the Census Bureau provided individual-level data from the 1940 Census on people of Japanese ancestry living in Washington, D.C. The information revealed included names, addresses, sex, age, and occupation. This information was apparently requested by the Secret Service because of a threat, the nature of which is not known, against President Franklin D. Roosevelt (Seltzer and Anderson 2007). For many years, census professionals denied that any records of Japanese Americans had been released (Gatewood 2001, p. 14–16).[8] It was a point of pride, used to bolster confidence in the Census Bureau, that even during wartime the data were kept confidential. When evidence of the release was discovered, Census Bureau professionals noted that the release of the information was legal in 1942. The conclusion many people could draw from this episode, however, was that a law protecting confidentiality could be repealed and replaced by a totally different law.

In fact, however, the story of the censuses since 1940 has been one of increasing concern for and protection of confidentiality.

5.3 Increased Legal Protection for Confidentiality

Title 13 of the U.S. Code has consolidated the various provisions for confidentiality of census information. Since 1954, results cannot be shared with anyone for non-statistical purposes. Typically, additional safeguards have come about through

---

[7] See, "Events in the Chronological Development of Privacy and Confidentiality at the U.S. Census Bureau" in references.

[8] Gatewood documents the variety of statements from the Bureau between 1988 and 2000 about this episode. Minkel's (2007) article cites Margo Anderson and William Seltzer's discovery in 2000 of evidence to document the data sharing. Minkel opens with the sentence "Despite decades of denials, government records confirm that the U.S. Census Bureau provided ... names and addresses of Japanese-Americans ..." See, Minkel, J.R. (2007).

amendment of Title 13. For example, in 1962 Title 13 was amended to extend confidentiality even to the copies that companies had retained of their own returns in the economic censuses. In 1973 the Census Director lost discretion to release individual information to governors of states and leaders of municipalities. Amendments in 1976 incorporated parts of the Privacy Act of 1974 and restricted the ability of the Census Bureau to communicate individual information to other federal agencies. The E-Government Act of 2002 updated the requirements for non-disclosure for the digital age. Census data are not subject to requests under the Freedom of Information Act (FOIA).

Federal courts have upheld Title 13's protections for confidentiality. For example, the courts have held that even other government agencies cannot access census information. The Supreme Court ruled specifically that address lists are protected from release.[9]

Sometimes the enforcement of Title 13 has come about through executive action. In August of 1980 the Census Bureau stopped the FBI from reviewing individual forms in a case in Colorado Springs. The forms had been sought under a warrant and the issue being investigated was the falsification of census records. The forms were returned to the Census Bureau without being opened. This action was achieved by discussion between the heads of the two agencies, and not judicial intervention (U.S. Census Bureau 2003).[10]

Every employee of the Census Bureau swears an oath to maintain the confidentiality of census records both as to specific information on the forms and the records themselves. The wording of this oath is as follows: "I will not disclose any information contained in the schedules, lists, or statements obtained for or prepared by the Census Bureau to any person or persons either during or after employment." Note that this is a lifetime obligation, binding employees even after they leave the Census Bureau. Violations can be penalized by fines up to $250,000 and imprisonment up to five years or both.

5.3.1   Publicity Concerning Confidentiality

The Census Bureau makes considerable efforts to ensure that the population understands that although completing the census form is required by law, the census data are also confidential by law and can be used only for statistical purposes. There is a substantial collection of information on confidentiality at www.census.gov, and there are also individual flyers that can be used to educate people.[11]

---

[9] An amendment to Title 13 in 1994 permits the sharing of census address information with states and sub-state jurisdictions for the limited purpose of improving the lists that will be used in carrying out the census and surveys.

[10] See, "Census Confidentiality and Privacy: 1790–2002" in references.

[11] For an example of such a flyer, see U.S. Census Bureau (2019a). "The 2020 Census and Confidentiality."

This information is quite explicit that census data cannot be shared with law enforcement or used against the respondent in any way. In fact, the Census Bureau identifies specific agencies that cannot access census data: the Federal Bureau of Investigation (FBI), Central Intelligence Agency (CIA), Department of Homeland Security (DHS), and Immigration and Customs Enforcement (ICE). The Patriot Act does not override the protection of census data. In addition, the information specifies that the information is not accessible through discovery in litigation and cannot be used against a respondent in court. Census data cannot be subpoenaed. Again, the Census Bureau is quite explicit in mentioning specific worrisome possibilities: census data cannot be used for immigration enforcement nor for determining benefits eligibility.

Given the elevated profile of immigration issues generally, these reassurances seem intended to persuade people who might be illegally present or engaged in illegal activity to complete the census anyway.

5.3.2 Circumstances Under Which Individual Information Can Be Released

Census data are not confidential forever. The "72-year rule" is a legal requirement that forbids public release of census records for 72 years (Title 44, U.S. Code n.d.). After 72 years, the results become available for sale or rental in microfilm from the National Archives and Records Administration (U.S. Census Bureau 2000).[12] According to this rule, the 1940 Census records became available in 2012, and the 1950 Census records are expected to be released in 2022. These historical records are used by genealogists, historians, and people who are curious about their family background.

Records that are still closed because the 72 years have not yet elapsed can be accessed only by the individuals themselves, the individual's heirs, and legal representatives of the individual. A family could need access to census information, if for example there is no birth certificate but there is a need to establish a person's age for a pension.

5.4 Indirect Disclosure

The protections of Title 13 are intended to prohibit direct disclosure of information from individual census returns. There is another circumstance to which Census professionals are alert, called indirect disclosure. Indirect disclosure occurs when legally available census data (i.e., published) can nevertheless be analyzed in such a way as to identify information about an individual or household.

---

[12] See, "Availability of Census Records about Individuals" in references.

Here is a hypothetical example. Suppose that a small town has a single family with ten children. If a table were published that showed household size for this town, it might be possible to identify this single family from the tabulation. This is already an issue when the census data is available principally in published tables, as it was for most censuses. More recently, however, census data have been available online so that users can construct the tabulations or statistical analyses they wish to use. This capability greatly increases the possibility that a determined analyst could discover information about an individual. The Census Bureau has worked for many years to make it impossible for this to happen.

5.4.1   Early Concerns with Indirect Disclosure

As early as 1920 the Census Bureau would "eyeball" data tables for business censuses to manually hide uniquely identifying information (called suppression) or to combine categories into larger categories (called compression). By 1930 the Census Bureau limited publication of small-area data because of the risk of indirect disclosure. Legislation after the 1940 Census directed the Census Bureau to identify and hide information that would uniquely identify individuals. In the 1970 Census, the Census Bureau suppressed some whole tables to protect small-area data about people and housing. There were even more tables suppressed in 1980.

By 1990, however, the level of concern about indirect disclosure was growing and some additional safeguards were added. With income data, for example, both rounding and top-coding were used. Top-coding refers to a broader category that is used for the top of a distribution. With income there are a few very high values; if same-sized intervals of dollar values were used, there would be cells with very few entries, and therefore a risk of indirect disclosure. If instead the top-code was "$150 thousand dollars or higher," then many more entries would appear in that cell and the risk of disclosure would be reduced.

5.4.2   Digital Data Increase Indirect Exposure Risk

For many decades the Census Bureau published information in the form of tabulations at various geographic levels and classified by various demographic techniques. Census automation made available many more combinations of tables had earlier been easily available. Microdata files, called PUMS (Public-Use Microdata Samples) became available to users on magnetic tape for the 1960 and 1970 Censuses.[13]

---

[13] PUMS are also available for ACS, and a description of confidentiality procedures is available via the Census Bureau. See, U.S. Census. (2018a). "Confidentiality of PUMS." PUMS data have been made available for the 1940–1990 Censuses, facilitating useful historical studies of the population. The data are publicly available and described at https://sedac.ciesin.columbia.edu/data/set/acrp-public-use-microdata-samples

These tapes contained a random sample of actual long form census records, with names and addresses stripped off, and offered the users the opportunity to do many new kinds of analyses. For example, a researcher interested in immigrants from a particular country could analyze them and compare their occupations and income with native-born Americans. Multiple regression analysis and other statistical tests were feasible compared with using published tables only.

Giving users so much access to the data greatly increased the usefulness of the data for users with special business or research needs. This access also increased the potential for indirect exposure, and Census Bureau statisticians worked on additional means for suppressing, swapping out, top-coding, or in other ways editing the data so as to make indirect exposure less likely.

The first step taken is removing individually identifying information, such as names and addresses.[14] Detailed geographic data are also omitted, so no information is available for small towns, census tracts, or other small census geographic units. Some variables such as age and income are top-coded. Top-coding is used to protect statistical outliers. So, for example, the top 0.5% of all responses (or 3% of non-zero responses) would have the same value attached to a variable such as income. This could be the cut-off number (such as a certain income level) or the interpolated median value of the variable. A bottom-code might be used for a variable such as the date a housing unit was built. Some information may be recoded or rounded ("blurred").

Census statisticians look for housing units or individuals with unique characteristics and seek to protect their identity. Data swapping is one such technique, in which the unique characteristics of a household might be "swapped" with those of another household, while leaving other characteristics the same. Another method is noise injection, which alters the micro-data without altering the frequency totals in census tabulations.

By 2000, census results were published online. Although appropriately anonymized, these data aroused additional concerns about indirect exposure. There was an enormous benefit to the user community from the access, and a democratic appeal to making so accessible to the people the information for which the people had paid. Census Bureau statisticians continued to experiment with statistical methods that would make accessible census data that were "true" in a statistical sense, yet without putting exactly identifying information in the public sphere that would permit identification of a household or individual.

In 1790 the U.S. Census was open both in the sense that the statistical information was public, and also that the census returns themselves were public. Identification of households was possible for the cost of a stroll to the town square. By 2000, the

---

[14] For a more detailed discussion of these measures, see McKenna (2019). "Disclosure Avoidance Techniques Used for the 1960 Through 2010 Census." In each of these censuses somewhat different methods were used, some of which are not well documented. Other methods, such as data swapping, are deliberately not described in detail to the public. This collection of measures is sometimes described as "ad hoc" measures, in contrast to the differential privacy method to be described below.

U.S. Census was still open in the sense that the statistical information was public, and its online availability made it far more accessible to much of the population than ever before. But many measures had been taken to keep the identifying information about households inaccessible and unknowable.

5.5 Differential Privacy

As early as 2006 Census Bureau statisticians began to work on an algorithm-based approach to avoiding disclosure. This approach was considered to be vital in the era of increasing digital threats. A differentially private data set on commuting patterns was published in 2008. By 2017 the Census Bureau announced that differential privacy would be used for the 2020 Census. By 2025 differential privacy will also be used for the American Community Survey.

5.5.1 What Is Differential Privacy?

Few statistical terms would appear to be less intuitively comprehensible than "differential privacy." Ron Jarmin, the Deputy Director and COO of the Census Bureau, describes differential privacy, a method developed by Microsoft researchers, as the "gold standard for privacy protection in computer science and cryptography" (Jarmin 2019a). Differential privacy arose as a set of statistical techniques to deal with the increased risk to confidentiality in a digital era (Dwork 2006; Dwork and Roth 2014). Jarmin describes the procedure as injecting "controlled noise into the data in a manner that preserves the accuracy at higher levels of geography."

Consider the context in which the 2020 Census will be taken, compared with earlier censuses. There are many databases, often tied to commercial enterprises such as Facebook or Google, with varying levels of coverage, confidentiality, and data security. Enter a person's name into Google and the odds are high that at least one hit will result. For a person with modest public visibility, there will be multiple hits. Even a casual internet user could combine available sources to reveal a reasonable profile of an individual. Even if someone could in this way learn many things about an individual, it is important that the information held by the Census Bureau— even if it is the very same information—not be individually identified.

Besides the commercially available databases, there are also many public databases that are required to be open access, such as voter registration data, property tax rolls, the directory information for students at public universities, and so on. While perhaps not so easy as a Google search, an afternoon at the local city hall could also provide a substantial profile on an individual. It is important to emphasize that these databases, which are often run by the states and not the federal government, are not bound by Title 13. Quite to the contrary, they are intended to be available for public inspection.[15]

The Census Bureau can and does promise to protect the confidentiality of the data that it collects, and it can require oaths of its personnel, rigid review procedures for its publications, and a host of other safeguards. These protections assume that the Census Bureau has all of the data within its control. Today however the Census Bureau must contend with the data that are outside its firewall. Thus, because of matters entirely outside its control, the Census Bureau cannot guarantee the confidentiality of its data (Abowd 2017).

Jarmin (2018) describes the underlying issue in this way:

> Today, from published census data alone someone could potentially determine that in a specific block there is a specific person who is 32, male, white, married and a father of two, but that person's name would still be missing. However, that same person could, with access to the right outside databases, link the reconstructed census information with additional outside information and zero in on that person's identity. The process is called re-identification, and that threat has become more real with today's technology. And we are reacting to these changes.

The concern about re-identification is not totally hypothetical, although I am unaware of any successful effort by an outside person to crack census confidentiality. Some research has indicated that a small number of individuals could be identified in commercially available data based on information from a census microdata source (Ramachandran et al. 2012). Ramachandran and associates used a PUMS from the ACS to study one county in each of the states of California, Florida, and Texas. The PUMS data set had been sanitized by the Census Bureau prior to release. Nevertheless, there were 62 data points for each case and the researchers found 926 individuals with unique combinations of state, geographic area, age, and sex. They reported identifying 87 of these individuals in a commercially available data set that contained names and addresses, among other items. While this is a very small percentage of the total population at risk, and there could have been false matches among the 87, the feasibility of re-identification was demonstrated.[16]

In 2020 the Census Bureau will deploy differential privacy through its Disclosure Avoidance System (DAS). The Census Bureau used a prototype system to provide privacy protection for the data collected during the 2018 Census Test in Providence, RI. As Abowd and Garfinkel report, the "DAS ingested the 2018 Census Edited File, ran the differential privacy algorithms, and produced the Microdata Detail File (MDF). The MDF was then used to produce the tabulations ... released on April 15,

---

[15] States vary in whether exceptions are allowed for privacy reasons—for example, for someone who is a stalking victim and has a protective order to conceal her location.

[16] A group of researchers at the Census Bureau reconstructed the record for individuals from aggregated census tables, and then attempted to match the reconstructed individuals to census block records. They found "putative matches" about half the time.

2019" (Abowd and Garfinkel 2019). The Census Bureau has released the source code for the DAS that was used in this test along with PUMS modules so that users can try out the privacy algorithms for themselves.

5.5.2   Is Differential Privacy Just Making Up Data?

The 2020 Census data transmitted to Congress for apportionment will be unedited. Apportionment is based solely on state of residence and population count, so privacy is not a risk. The numbers are too large, the geography is too high-level, and the variables too few for re-identification. Subsequent data releases, however, will have the differential privacy procedures applied.

Is the "noise injection" simply a way of making up information? There is an inevitable trade-off of privacy for accuracy of information. The Census Bureau is deciding that the loss of accuracy will be acceptable given the improvement in privacy. This decision has not received unanimous praise within the scientific community. Census Bureau personnel have made the rounds of various professional meetings and found consternation about the extent to which the accuracy of the data will be assured. Professor Steven Ruggles of the University of Minnesota, an outspoken critic of the new approach, says, "This is not the time to impose arbitrary and burdensome new rules that will sharply restrict or eliminate access to the nation's core data sources" (Mervis 2019).

The Census Bureau hopes that making the process of differential privacy transparent will maintain confidence that the data are being handled correctly and that the results are accurate. Data users outside the Census Bureau respond that noisier data will ripple into the many other uses of census data. And more stringent limitations on the public use of the 2020 Census microdata may be one of the ripples. Finally, it is argued that differential privacy is a complicated concept and the Census Bureau needs to do more to communicate it to the public. Otherwise, a measure adopted in part to reassure the public about the confidentiality of census data may instead confuse the public and breed more mistrust.

5.6   Data Safety and Hacking

In its efforts to reassure the population that the data are safe, the Census Bureau has referenced its partnership with Microsoft and the other safeguards it has undertaken to protect the data (Jarmin 2019a, b). Nevertheless, the fact that Americans will be invited to use the internet as the preferred mode of response has aroused concerns.

5.6.1   *Accessibility to the Internet*

One concern of course is accessibility. Not every household has access to the internet, and the people without good access are often those who are already in hard-to-count populations, such as minority group members, low-income households, people with housing instability, and rural residents. As described in Chap. 3, the Census Bureau is already preparing to provide mail-back questionnaires in certain neighborhoods where accessibility is expected to be a problem. And census workers will be concentrating their efforts on the hard-to-count population.

The growth of accessibility to the internet in the United States has been astonishing. Smart phones have provided important access to many people who previously did not have it, and schools, libraries, recreation centers, and other locations can amplify access. Public wi-fi is widely available. But are these public computers and wi-fi networks safe from hacking and other cyber-mischief?

5.6.2   *Cyberthreats*

Directly addressing the issue, the Census Bureau says:

> Our IT systems are designed to defend against and contain cyberthreats. From the beginning of the data collection process through the final storage of information, we protect your data following industry best practices and federal requirements. We use data encryption and two forms of authentication to secure system access. The security of our systems is a top priority, and we continually refine our approach to address emerging threats and position ourselves to identify, prevent, detect, respond to and recover from possible cyberthreats. (The U.S. Census Bureau 2018b).[17]

Microsoft's Defending Democracy program started as an effort to make elections more secure as electronic voting systems became more popular. The 2020 Census is an additional institution of democracy participating in this program.

A major thrust of this effort has been preventing hacking through end-to-end encryption. Necessarily, however, this effort must be addressed to what is within the government's control. The government can seek to control its devices, networks, and storage. As with differential privacy, where there is information the government does not control, with hacking there are failures of hardware and malware that the Census Bureau may not be able to correct. Suppose, for example, that a computer virus became widespread that could identify keystrokes from the smartphone as the respondent enters it, and before the encryption can occur.

A larger concern is whether the protection against cyberthreats will be robust enough to protect against a major data breach such as the one that affected the Office of Personnel Management.

---

[17] See, "Data Stewardship" in references.

In addition, the Census Bureau is concerned about misinformation and disinformation about the census, especially as these could affect the willingness of the population to complete their census return. One could imagine various phishing schemes designed to send respondents to an impostor site rather than to the Census Bureau. Website searches about the census on Microsoft's search engine Bing will place a graphic at the top of the page to direct questioners to verified information. There is also an effort to identify and address fake websites that try to imitate the genuine website, which is www.census.gov. Facebook is collaborating with the Census Bureau to promote participation in the Census (Jarmin 2019b, Defending Democracy).

Finally, given that readily available, accurate information will be important, the Census Bureau has unveiled a variety of approaches to connect with teachers, journalists, community leaders, and others with classroom activities, story ideas, and tidbits about the benefits that census participation can bring to the locality.

5.7   Summary

For over 150 years, the U.S. Census has been collected with increased attention to confidentiality and privacy. The rapid development of digital technology and the explosion of digital data of various types is a challenge for the 2020 Census. The 2020 Census will be the first for which internet response will be the preferred mode of response. Reassuring the public that their data will be confidential is challenging when various cyberthreats and data breaches are reported every day.

To better ensure that census data cannot be used for reidentification of a respondent or a household, the Bureau has announced that an algorithmic approach to data security called differential privacy will be used in 2020. Some members of the scientific community have expressed skepticism about this move, the significance of the risk, and the Bureau's position on the trade-off of accuracy for privacy.

Besides the issues of formal privacy, there are also concerns that census data may be compromised. The Census Bureau has announced a partnership with the Defending Democracy program at Microsoft, and a number of measures are underway to protect the collection and storage of the data. The Census Bureau is also concerned with possible misinformation or disinformation being disseminated about the census. A robust communication plan to provide verified information is underway.

References

Abowd JM (2017) How will statistical agencies operate when all data are private? J Priv Confid 7(3):1–15

Abowd JM, Garfinkel SS (2019) Disclosure avoidance and the 2018 census test: release of the source code. https://www.census.gov/newsroom/blogs/research-matters/2019/06/disclosure_avoidance.html. Accessed 10 Dec 2019

Dwork C (2006) Differential privacy. 33rd International Colloquium on Automata, Languages and Programming, part II (ICALP 2006), Springer Verlag, 4052, 1–12, ISBN 3-540-35907-9

Dwork C, Roth A (2014) The algorithmic foundations of differential privacy. Found and Trends Theor Comput Sci 9(3–4):211–407. https://doi.org/10.1561/0400000042

Fay RL, Carter W, Dowd K (1991) Multiple causes of nonresponse: analysis of the survey of 1990 census participation. Proceedings of the Social Statistics Section. Alexandria, VA: American Statistical Association

Gatewood G (2001) A monograph on confidentiality and privacy in the U.S. Census. https://www.census.gov/history/pdf/ConfidentialityMonograph.pdf. Accessed 10 Dec 2019

Hill JM, Rothhaas CA, Lestna FA, Dusch GS (2012) 2010 confidentiality notification experiment report. Report CPEX-170. https://www.census.gov/programs-surveys/decennial-census/decade/2010/program-management/cpex/2010-cpex-170.html. Accessed 8 Dec 2019

Jarmin R (2018) The balancing act of producing accurate and confidential statistics. https://www.census.gov/newsroom/blogs/director/2018/12/the_balancing_actof.html Accessed 10 Dec 2019

Jarmin R (2019a) Census bureau adopts cutting edge privacy protections for 2020 census. https://www.census.gov/newsroom/blogs/random-samplings/2019/02/census_bureau_adopts.html. Accessed 19 Dec 2019

Jarmin R (2019b) Defending democracy. https://www.census.gov/newsroom/blogs/random-samplings/2019/07/defending_democracy.html. Accessed 16 Dec 2019

McKenna L (2019) Disclosure avoidance techniques used for the 1960 through 2010 census. CED-DA Report Series 2019–07. https://www.census.gov/library/working-papers/2019/adrm/six-decennial-censuses-da.html. Accessed 6 Dec 2019

Mervis J (2019) Can a set of equations keep U.S. Census Data private? Science (January 17). https://doi.org/10.1126/science.aaw5470

Minkel JR (2007) Confirmed: the U.S. Census Bureau gave up names of Japanese-Americans in World War II. Scientific American. March 30. Https://www.scientificamerican.com/article/confirmed-the-us-census-b/. Accessed 3 Dec 2019

Naylor B (2016) One year after OPM Data breech, what has the government learned? NPR, June 6, 2016. Retrieved from https://www.npr.org/sections/alltechconsidered/2016/06/06/480968999/one-year-after-opm-data-breach-what-has-the-government-learned. Accessed 19 Dec 2019

Ramachandran A, Singh L, Porter E, Nagle F (2012) Exploring re-identification risks in public domains. https://www.census.gov/content/dam/Census/library/working-papers/2012/adrm/rrs2012-13.pdf. Accessed 12 Dec 2019

Seltzer W, Anderson M (2007) Census confidentiality under the second war powers act (1942–1947. Paper prepared for the annual meeting of the population Association of America. New York, March 30

Singer E, Mathiowetz NA, Cooper MP (1993) The impact of privacy and confidentiality concerns on survey participation: the case of the 1990 U.S. Census. Public Opinion Quarterly 57 4(Winter):465–482. https://doi.org/10.1086/269391

Sullivan T (2020) Coming to our census: how social statistics underpin our democracy (and Republic). Harvard Data Sc Rev 3. 1 (January) forthcoming

The U.S. Census Bureau (2000) Availability of census records about individuals. https://www.census.gov/prod/2000pubs/cff-2.pdf. Accessed 13 Dec 2019

The U.S. Census Bureau (2003) Census confidentiality and privacy: 1790-2002. PP 18–20. https://www.census.gov/library/publications/2003/comm/monograph-confidentiality-privacy.html. Accessed 9 Dec 2019

The U.S. Census Bureau (2018a) Confidentiality of PUMS. https://www.census.gov/programs-surveys/acs/technical-documentation/pums/confidentiality.html. Accessed 7 Dec 2019

The U.S. Census Bureau (2018b) Data Stewardship. https://www.census.gov/about/policies/privacy/data_stewardship.html. Accessed 15 Dec 2019

The U.S. Census Bureau (2019a) The 2020 census and confidentiality. https://www.census.gov/content/dam/Census/library/factsheets/2019/comm/2020-confidentiality-factsheet.pdf. Accessed 10 Dec 2019

The U.S. Census Bureau (2019b) A history of census privacy protections. https://www2.census.gov/library/visualizations/2019/communications/history-privacy-protection.pdf. Accessed 8 Dec 2019

The U.S. Census Bureau (n.d.) Events in the Chronological Development of Privacy and Confidentiality at the U.S. Census Bureau. https://www.census.gov/history/pdf/PrivConfidChrono.pdf. Accessed 7 Dec 2019

U.S. Code (n.d.) Title 44

Wolff J (2019) Cybersecurity experts are leaving the Federal Government. That's a Problem. New York Times. https://www.nytimes.com/2019/12/19/opinion/cybersecurity-departures-government.html. Accessed 19 Dec 2019

Chapter 6
Are There Alternatives to the Census?

Abstract The continuous population register is used in some countries to track population composition, change, and movement. Although confidentiality is a concern with registers, the population registers are more current than a census and permit applications not possible with a census. The United States does not currently have a federal-level continuous population register, but there are both federal and state databases that function as partial population registers. These databases are administrative records, collected for one purpose but potentially useful for supplementing census data. In the 2020 Census, administrative records will be used both for data checking and for completing some item non-response. By Executive Order President Trump has ordered the Census Bureau to produce a database of voting-age citizens using federal administrative records to create a "best-citizenship" variable. Creating this database requires statistical modeling to create the estimate of citizenship and research on how to match names and addresses from the census to the administrative records. Under Title 13 this database will be confidential but will also provide state legislatures a way to redistrict based only on citizens of voting age. If successful, this database will raise the issue of whether a continuous population register would be possible for the United States.

Keywords Administrative record · Continuous register · Partial register · Record linkage · Citizenship · Confidentiality

Sticking to traditional approaches within the demographic research community might prevent further progress, or just let other, bolder, communities of scholars bring the advances needed to further our understanding of population processes ...A fruitful way ahead is perhaps to combine traditional approaches with new one: counting and now-casting, indirect estimation and the used of non-representative Web-based data, official statistics and digital breadcrumbs. (Billari and Zagheni 2017: 176).

As important as the U.S. Census is, and as significant as its Constitutional mandate is, the census is but one of multiple data sets available today to government officials, business leaders, community organizers, teachers, and the general public. Chapter 5 reviewed the dangers that the proliferating data sources may pose in terms of allowing individual reidentification of someone included in the census. These proliferating data bases offer a totally different possibility as well: replacing the decennial census with another approach or by combining different approaches.[1]

6.1   The Continuous Population Register

One alternative to the census, the continuous population register, is used in some countries, such as those in Scandinavia (Poston Jr. and Bouvier 2017, p. 46–54). The population register tracks every member of a population from birth until death. Although countries vary somewhat in the information that they register, some typical possibilities include school attendance and school leaving; address and changes of addresses; military service; marriages and divorces; and eligibility for and receipt of government benefit programs. Depending upon the country, eligibility for health care, pensions, further education benefits, and other programs might be included.

The Chinese tradition of registering population dates back to the Han dynasty (206 BC – 220 AD) and was adopted elsewhere in Asia (Taeuber 1959, p. 261). In Europe the continuous population register originated before modernity in religious parish records (Shryock et al. 1976, p.13). Later the population registers, sometimes termed civil registers, were maintained by local and then national governments.

Sweden forms an exemplar of this change from religious to secular record-keeping. By royal decree, births and deaths have been recorded in Sweden at least since the 1600s. Until 1991 the Church of Sweden (and later congregations of other faiths) maintained the register, and since then it has been maintained by the Swedish Tax Authority. Any change in residence longer than six months must be recorded, along with place of birth, citizenship, immigration to or emigration from Sweden, and a personal identification number (PIN). This PIN is used for most interactions with the government. Individuals in Sweden have the legal right to see

---

[1] Eliminating the census would require a constitutional amendment, and that is a lengthy process. U.S. Constitution, Art. V. On Big Data alternatives to the census, see Billari, F. C., & Zagheni, (2017).

any information contained about themselves in the registry. Because of the length of time this registry has been kept, it has multiple medical, historical, and demographic uses.[2]

6.1.1   Advantages of a Continuous Population Register

A census is a snapshot of a country at a specific point in time. In the United States, this snapshot is available only once a decade. In reality, however, populations are always in motion: people within the population are continually being born, dying, and moving from one location to another. The population register better represents this dynamism. Moreover, because the register is always being updated, there is no need to ramp up every decade to take the census. The cost of maintaining the register could well be substantially less than the cost of planning, testing, advertising, conducting, and then analyzing the census. The need for parallel agencies, such as a census bureau and vital statistics bureaus, could be reduced.

6.1.2   Disadvantages of a Continuous Population Register

To be sure, the population register is not free of error. Potentially an illegal alien could avoid inclusion in the register, and individuals within the population could deliberately or forgetfully fail to register every event when required. While a single system is efficient, the existence of parallel systems (e.g., census and vital statistics) permits the developments of quality checks of the data.

The multitude of governments within the United States is also potentially an issue. Would a population register be the province of the federal government, which is currently responsible for the census, or of the fifty-plus vital records bureaus now run by the states and the District of Columbia? And if the latter, how would coordination be achieved?

Population registers need to account for migration, and the United States is a very mobile society. Americans are used to the idea of registering births and deaths with their states, but a requirement to register every move from one apartment to another would not easily fit Americans' mobile lifestyles.

Moreover, having a single PIN that could be used in every government transaction could seem an invitation to cyberthreats. The Social Security number does appear to function as a single PIN for many government functions and hacking of Social Security numbers has been a major cyberthreat.

---

[2] For example, see Barklay, K.J. & Kolk, M. (2018).

6.2   Partial Population Registers

The United States has many administrative records that function as partial population registers. An administrative record consists of data collected for a particular governmental purpose, such as conducting a particular program. Unlike the continuous population register, these administrative records are partial registers because they do not cover the entire population. And unlike Title 13 data, which are collected solely for statistical purposes, the administrative data are originally collected for a different purpose but might become useful for statistical purposes.

An example of a partial population register is driver's license records. The states issue driver's licenses to eligible individuals, typically people over a minimum age who have passed required tests and paid a fee. The principal purpose of the driver's license is to show that a person has the appropriate qualifications to drive a motor vehicle. Driver's licenses have many other uses beyond those of the Motor Vehicle Department. The driver's license is used as a form of identification, and it is sometimes used to register to vote, to apply for a job, and even to indicate an intent to become an organ donor. The registry of driver's licenses in many states covers a large fraction of the population of driving age.

Many federal administrative records cover a large enough portion of the population to qualify as partial registers. The Internal Revenue Service has tax records for most adult Americans and many children. Social Security, especially Medicare, is believed to have nearly universal coverage of the senior citizen population. Should Medicare-for-all become a public policy, then there is a possibility that Medicare could function as a continuous population register. Selective Service is a partial register for men over 18 years of age, although it is not kept up to date for men older than the age at which the former military draft operated. The Veterans Administration has data on veterans and their service records.

Other partial registries are kept by the states. Some examples include state tax records, land ownership records, motor vehicle registries, hunting and fishing licenses, and voting rolls. In some states there are also registries for certain occupational licenses, school attenders, concealed handgun permits, convicted sex offenders, and recipients of various benefits such as Medicaid. Unlike confidential Title 13 data, some state registers are required to be publicly available under the state's Freedom of Information Act (FOIA).

There are also non-governmental records that might be considered partial population registers. Credit agencies, although private, cover a large fraction of the population, and include information on name, address, income, occupation, Social Security number, and sometimes co-borrowers (often a spouse), in addition to the expected information on the status of credit accounts and pending legal actions. It seems increasingly likely that as social media cover more of the population there will be more ways in which a Facebook or Twitter account might be "scraped" as partial registries.

Given the discussion of differential privacy in Chap. 5, the reader has probably already anticipated that there are many ways to misuse administrative data and that some safeguards are necessary.

The regulation of Electronic Medical Records (EMRs) is instructive. An EMR assembles information from the various medical visits and hospitalizations of a patient. Current medications, test results, and diagnoses are included. A health care provider has immediately at hand the patient's medical history and can add symptoms, vital signs, and other information. This information saves time, minimizes prescription errors, and generally provides better care. For a patient who is brought unconscious to a hospital, the EMR can be a lifesaver.

On the other hand, the EMR takes some of the most sensitive information about a person and combines it into an electronic file that could potentially be hacked or otherwise misused. To protect medical information, Congress has passed a stringent act called HIPAA (Health Insurance Portability and Accountability Act). Among other things, HIPAA establishes industry-wide standards for electronic billing and other health care information, and it requires the protection and confidential handling of this information.

6.2.1 Linking the Partial Registers

Given the care and effort devoted to safeguarding this one type of medical information, and the issues around privacy and confidentiality discussed for census data in Chap. 5, it is apparent that the linking of registers is fraught with issues. Linking records is not technically difficult. In fact, it is so easy to do that there are laws that limit the linkage of certain records, that specify the permissions required for linkage, or that list the procedures that must be followed to allow the linkage.[3] Why this linkage is significant for Census Bureau operations is discussed below.

6.3 Census Bureau Use of Administrative Records

The Census Bureau accesses the federal partial population registers and other administrative data for many of its functions. In fact, Title 13 of the U.S. Code authorizes the use of administrative data instead of direct inquiries "to the maximum extent possible with the kind, timeliness, quality, and scope of the statistics required." Because of this authorization the Census Bureau is proficient at accessing

---

[3] It is probably not a coincidence that in the Scandinavian countries, where so many pieces of information concerning an individual can be linked, there are also some of the most stringent confidentiality and privacy provisions.

the records of other federal agencies. This is, by the way, a one-way street: information can come from a federal agency to the Census Bureau, but that same agency cannot requisition census records.

Congress has expressed a wish to reduce the respondent burden or survey fatigue that comes from repeated questioning. Respondent burden is often cited as a reason for shorter questionnaires, for paperwork reduction requirements, and for requiring government documents to be written in plain English. The development of the American Community Survey and the retirement of the census long form represent a desire to reduce the burden on respondents while having more timely information. Filling in the answers to census or survey questions with answers that are already in an administrative record is cost-effective and requires less time and effort of the respondent (Ortman 2018).

The average American may feel besieged from the sheer volume of commercial messages, robo-calls, solicitations, and surveys from non-governmental organizations. While this deluge of requests is not the fault of the Census Bureau, the Census Bureau must nevertheless deal with the declining response rates that result when Americans are simply fed up with answering one more questionnaire.

As Chap. 3 indicated, as part of non-response follow-up in the 2020 Census, the Census Bureau is testing the use of administrative records to impute race, age, and Hispanic origin if these pieces of information are missing. Information from administrative records will also be compared with census returns as an error check. More extensive use of administrative records with the ACS is also planned. And for many years the Census Bureau has used administrative records such as birth and death records to make estimates and projections of population size between censuses.

6.3.1 Not All Information Is Treated the Same Way

Federal agencies participate in information sharing programs that are limited by their statutory authority. Non-confidential information is shared among agencies that face similar issues, such as hacking or cybersecurity problems and solutions (Rockwelll 2017). An important fact to reinforce is the one-way nature of the linkage of individually identifiable data to the Census Bureau. Because of Title 13, data about households or individuals that comes into the Census Bureau also comes under the shield of confidentiality. This means that the census information can be used only for statistical purposes.

6.4 Developing a Citizenship Count (CVAP)

With this background on registers, administrative records, and linkages, we return to the issue of citizenship and the 2020 Census. As explained in Chap. 4, the Supreme Court struck down the addition to the 2020 Census of a question about citizenship.

About two weeks after the Supreme Court's decision, President Donald J. Trump issued an executive order instructing federal departments and agencies to provide to the Census Bureau the citizenship data that they already held in their databases. President Trump said, "Some states may want to draw state and local legislative districts, based upon the voter eligible population" (Rogers et al. 2019).

The first release of data from the 2020 Census will be Census Unedited File, which is the population counts by state, scheduled for preparation by November 30, 2020, and release to the President and Congress by December 31, 2020. What will be contained in this file is the count of the population for each state, together with overseas federal employees and their dependents. That overseas population will be allocated back to its state of residence. These data are then analyzed with the apportionment formula to produce the number of representatives allocated to each state. The CUF does not contain any citizenship data.

Redistricting data at the block level will be produced in the Census Edited File, which will be released state by state between February 18, 2021, and March 31, 2021. The Census Edited File differs from the Census Unedited File because it has used administrative records and some statistical modeling for imputing missing values. The Census Edited File goes through the Disclosure Avoidance System to minimize the possibility of identifying any individual in the data.

The President's Executive Order 13880 commits the Census Bureau to release the Citizen Voting-Age Population (CVAP) data by March 31, 2021 (U.S. Census Bureau 2019).[4] This is basically the same time frame as the release of the Census Edited File (CEF). CVAP will combine administrative data from a number of federal agencies into a separate microdata file that will contain a "best citizenship" variable for every person in the 2020 Census. The Disclosure Avoidance System will be used with this microdata file. The same confidentiality rules that apply to the CEF will also apply to the citizenship variable. The data will be produced at the block level and will be available to the public.

6.4.1 *"Best Citizenship" Variable*

There is an internal working group in the Census Bureau that will release the specifications for the CVAP by March 31, 2020. The Census Director has convened the Interagency Working Group, consisting of high-level executives in federal agencies whose databases have person-level data relevant to estimating citizenship. To be useful the administrative records will need to have variables that contain citizenship data and that can link to names and addresses on the census record.

Among the possible sources of citizenship data that the Census Bureau will examine are the following (U.S. Census Bureau 2019)[5]:

---

[4] See, "Citizen Voting-Age Population Data" in references.

[5] See, "Citizen Voting-Age Population Data" in references.

- Social Security Administration NUMIDENT, which contains place of birth and citizenship status for approximately 94% of the population
- Internal Revenue Service 1040 and 1099 forms, which would be used for purposes of most current address
- CMS Medicare and Medicaid/CHIP, which contain some citizenship data but are needed for current address
- Housing and Urban Development, which potentially contains current address information from Federal Housing Administration, Public and Indian Housing Information Center, Tenant and Rental Assistance Certification System, Low-income Housing Tax Credits, and Computerized Homes Underwriting Management System
- Department of Homeland Security USCIS/CBP/ICE, which contains information on lawful permanent residents and naturalization data (CIS), visas (ICE), arrival/departure (CBP)
- Department of State (Passport Services), for citizenship data from passports
- Social Security Administration, for information from the master beneficiary record
- Indian Health Service, for patient registration data
- Department of Justice, for US Marshals and Citizenship and Immigration Data Collection

An important part of the research program at the Census Bureau will be the statistical modeling needed to combine citizenship data from various sources to produce the "best citizenship" variable. This variable is the "best" in terms of providing the best estimate of whether someone is or is not a citizen. Some examples of variables that could go into this model are place of birth, naturalization data, and passport information.

6.4.2 Linking Databases

A second important research effort will be improving the linkage of names and addresses from these different databases. There are many issues here: people with the same name, people who have had many addresses, people who moved just before the census (and before they notified Social Security and other agencies of the address change). And the process will need to be repeated for millions of people over the age of 18.

The Census Bureau is still studying the linkage issues, and so what appears below is conceptual guesswork about how the process could work. Suppose the issue is whether the next person in the census count should receive a best-citizenship designation as "citizen" or "non-citizen." Let us hypothesize that this person is named John Smith and the census return shows that his household resides at 1234 Main Street in Wallbridge, Texas. There are many John Smiths in the various data files, so the computer searches for Main Street and Wallbridge, Texas. If there is a match—let's say with Social Security records—and it shows the same name and address, and a birthplace in Tyler, Texas, then presumptively John Smith can be designated a citizen. People born in the United States are U.S. citizens.

Suppose on the other hand, that John Smith was born in Germany, where his father was stationed on military service. Children born abroad to U.S. citizens are also citizens, but discovering whether John Smith qualifies may be more complicated. If John Smith holds a passport, that will settle the issue by linking to the passport data. If John Smith does not hold a passport, then additional searching through the other government databases will be necessary. He will not, for example, be a visa-holder because he is a citizen.

Suppose alternatively that the Social Security record shows many John Smiths, none of whom is currently living in Wallbridge. Perhaps an IRS record will prove a match for the Main Street address in Wallbridge and provide a Social Security number. It is very likely that John Smith uses his Social Security number as his Taxpayer Identification Number. That number could link into the Social Security information and to show that John's birthplace is in Texas. A matching program will need to be developed that could potentially go through 3, 5, 7 or more databases before finding a name and address match, and then go back through the databases looking for citizenship data.

The most difficult cases will be for the truly undocumented, for whom there might be no more information than the census record. For the person who is truly undocumented in any other government database, the issue will then be whether this person is presumptively a non-citizen, or simply someone for whom the matching algorithm failed.

Given its history of seeking to identify errors and improve data, the Census Bureau will almost surely launch an effort to evaluate the success of the matching program, particularly in identifying false positives (non-citizens labeled citizens) and false negatives (citizens labeled as non-citizens). The number of unsuccessful matches is likely to vary from state to state. For a state with a large net migration, such as Florida or Texas, the address matching could be particularly problematic. A large mismatch rate, in turn, would raise concerns about the fairness of redistricting using the CVAP. It is difficult in advance to determine the data quality issues that could arise in a database that has not yet been produced.

The legal issues around whether the legislatures may then redistrict using CVAP are not entirely clear, although the *Rucho* decision in 2019 might mean that the courts will take a hands-off position.[6] Almost surely there will be legal challenges before state legislatures complete redistricting with the new "best citizenship" variable.

---

[6] See Chap. 4 for a discussion of *Rucho v. Common Cause*, 2019

6.5   Do We Still Need a Census?

If the Census Bureau can correct census data with administrative records, and if the Census Bureau can successfully create a best-citizenship variable constructed entirely from administrative records, then it appears to be feasible for the Census Bureau to construct a database that appears similar to a continuous population register.

Such a continuous register would require the cooperation of vital statistics bureaus, located in the states, to document when someone enters the continuous register through birth and when someone leaves the register through death. The Department of Homeland Security could provide information on legal entrants to the United States and on legal entrants who leave the United States.

Problems would remain that have to do principally with the mobility of the population. U.S. citizens who emigrate to other countries, while typically relatively few in number, are harder to count unless, perhaps, they are receiving a Social Security check abroad. Internal migration will raise the issue of permanent addresses for the register. In particular young people move a lot as they go to college, join the military, find and change jobs. Then they often change residences as they progress through the family life cycle. Similarly, retirees are often a mobile population, sometimes having both summer and winter homes or downsizing and moving to be near adult children.

And the problem of illegal entrants would remain difficult. A foreign national who enters the United States undetected would probably not be included in the register. A foreign national who enters on a tourist visa and then overstays the visa could potentially be identified, although perhaps without a useable address.

The geographic location issues are critical if something like a continuous population register were to replace the census. Even if the Constitution were to be amended to eliminate the census in favor of a continuous population register, there would still be a need to reapportion the House of Representatives and to redistrict the states.

The CVAP will be an important development to answering some of these questions, even if it is never used to redistrict a single state. A successful CVAP will represent the feasibility of linkages among government databases to account for every resident in the country. With the addition of successful federal-state linkages—an addition that presents both legal and technical challenges—the CVAP would represent a prototype continuous population register.

And assuming that Title 13 would remain in effect and would apply to CVAP or whatever population register would replace the census, then the confidentiality of the information would be legally assured.

6.6   Summary

Censuses are not the only way to provide statistics about the population. Population registers have been used successfully in many parts of the world. The United States has administrative records that were designed for specific purposes, but that nevertheless have potential to serve as partial population registers. A number of federal agencies have such databases, and the states have their own partial registers.

President Trump's Executive Order to produce a citizenship voting-age population (CVAP) file to be provided to the states offers two challenges to the Census Bureau. The first is to develop a "best-citizenship" variable relying on administrative records rather than asking a citizenship question in the census. The second is to develop successful linkages and record matches with many federal databases.

Assuming that the CVAP process is successful and that the mismatch error rate is considered tolerable, then these developments raise the issues of whether the United States can and should develop a federal continuous population register. Such a register would require additional record linkages to the states' vital statistics bureaus. It would be subject to its own errors and shortcomings. Most importantly, for the continuous population register to supplant the census, an amendment to the United States Constitution would be required.

References

Barklay KJ, Kolk M (2018) Birth intervals and health in adulthood: a comparison of siblings using Swedish register data. Demography 55(3):9290955

Billari FC, Zagheni E (2017) Big data and population processes: a revolution? https://doi.org/10.31235/osf.io/f9vzp. Accessed 19 Dec 2019

Ortman J (2018). Survey fatigue, time crunch may have lowered response. https://www.census.gov/library/stories/2018/11/administrative-records-offset-declining-census-survey-response.html. Accessed 19 Dec 2019

Poston DL Jr, Bouvier L (2017) Population and society, 2nd edn. Cambridge University Press, Cambridge/New York

Rockwelll M (2017) Information sharing is complicated, even inside the government. https://fcw.com/articles/2017/04/06/cyber-info-sharing-rockwell.aspx. Accessed 20 Dec 2019

Rogers K, Liptak A, Crowley M, Wines M (2019) Trump says he will seek citizenship information from existing Federal records, not the Census. New York Times (July 11). https://www.nytimes.com/2019/07/11/us/politics/census-executive-action.html. Accessed 19 Dec 2019

Shryock HS, Siegel JS et al (1976) The methods and materials of demography. Academic Press, New York

Taeuber I (1959) Demographic research in the Pacific area. In: Hauser PM, Duncan OD (eds) The study of population. University of Chicago Press, Chicago, IL, pp 259–285

U.S. Census (2019) Citizen voting-age population data. https://www2.census.gov/cac/sac/meetings/2019-09/update-disclosure-avoidance-administrative-data.pdf?#. Accessed 19 Dec 2019

Postscript: Integrity and Partisanship

> The instability, injustice, and confusion introduced into the public councils, have, in truth, been the mortal diseases under which popular governments have everywhere perished; as they continue to be the favorite and fruitful topics from which the adversaries to liberty derive their most specious declamations.—James Madison[1]

James Madison, who played such an important role in framing the questions for the first census, also understood the inevitability of parties and partisanship. The use of the census for reapportionment guaranteed that every census would produce winners and losers, and therefore that every census would generate some level of political heat. The issue in 2020 as in 1790 is conducting an accurate census without turning the census into a political football.

There are many ways for partisanship to affect the census, including the political appointees who direct the Census Bureau and its home in the Department of Commerce. In addition, Congress can grant or withhold funds or change the requirements. The states and localities are also active, certainly in advocating before the census for a complete count and in some cases suing the federal government after the census to complain about the outcome.

In the case of the 2020 Census, the inflamed political atmosphere over immigration, nationality, and race (and Islam) helps to create an atmosphere of fear and suspicion that can affect the count. The fearful may avoid the count, and the skeptical may encourage them to avoid being counted. The efforts of the Census Bureau to be accommodating, such as providing the forms in thirteen languages, may have the effect of aggravating either those who advocate English-only for all government documents or those who claim that a fourteenth language should have been added.

Madison's answer to the dangers of factionalism (today we would say partisanship) was the system of checks and balances set up in the Constitution. The President's appointments required Senate confirmation. Monetary appropriations had to originate in the House of Representatives. The President could veto

---

[1] Madison, J. (1787). "The Same Subject Continued: The Union as a Safeguard Against Domestic Faction and Insurrection." *Federalist Papers* No. 10. https://www.congress.gov/resources/display/content/The+Federalist+Papers#TheFederalistPapers-10

T. A. Sullivan, *Census 2020*, https://doi.org/10.1007/978-3-030-40578-6

Congressional bills, but Congress could override the veto. In extremity, Congress had the power of impeachment. The Senate enjoyed longer terms and could afford to take a longer view than the House of Representatives. The judiciary interpreted law and served as a check to prevent the other two branches from acting unconstitutionally.

Madison also correctly foresaw that the fledgling government would need information in many areas. Statistics as a discipline was in its infancy, but its worth was quickly perceived. Although the census never asked about all the items that Madison proposed in his day, today the government collects data on everything from the aerospace industry to zoological parks. Data became an important commodity in and of itself, and many industries and localities in the United States benefited over time from the provision of free public data.[2]

It has always been a temptation for the party that holds the White House to seek to shade government statistics so that the current administration looks better or perhaps so that previous administrations look worse. This happens with many statistics that are generated both within the Census Bureau and other agencies, such as the unemployment rate, the poverty rate, or the undercount rate. These seemingly dry and lifeless statistics become imbued with emotion when they are depicted as one party's callous neglect of the poor and unemployed or of another party's search for votes by encouraging grievance and division.

The attack on the statistics ("fake news") has too often turned into an attack on those who produce the statistics ("deep state").

I want to take this opportunity to defend the producers of the statistics, who are far from being saboteurs of democracy and are instead among its greatest servants. I have never worked at the Census Bureau, but I have had the opportunity in my professional career to meet many of the data experts. The census professional staff includes statisticians, demographers, economists, geographers, computer scientists, and other highly trained people who pursue the best possible census. They have been committed to the non-partisan nature of the Census Bureau. I certainly do not know all of them, and I am forming a generalization on them without having anything approaching a random sample. Still, from my experience the census professionals have been doggedly determined to do the very best job they know how, to explain the census to the public as best they can, and to listen patiently to complaints, disagreements, and arguments.

Without doubt, the U.S. Census Bureau has been an exemplar to the world as a leader in many aspects of data design and analysis. They created many of the most significant innovations in survey technology, computer applications, and statistical analysis that are used worldwide. They have been alert to new issues, such as cyberthreats, as well as the long-standing issues such as the hard-to-count population. They are relentless in seeking to improve their methods, publicizing their

---

[2] The provision of online data has made much of the government's statistical data available for free, although published data may still be priced to recover the cost of production, and special tabulations for a user require fees. While not all the data are free, the cost is very low, especially in consideration of the cost to collect the data.

results so that everyone can know what data problems have arisen and how they have been remedied.

Within the professional community, there is often a vigorous argument over the innovations. For several decades there was a great debate over the proper formula to use for apportionment.[3] There was a debate in the 1980s and 1990s over the use of sampling to adjust the undercount, which finally ended up in the Supreme Court. Editing and imputing data have been controversial. The introduction of differential privacy has already sparked lively discussion about whether it is a solution to a phantom problem and a solution that does some damage.

All these arguments should be understood for what they are: the way that scientists seek to find truth in a real world that is usually ambiguous and defies categorization. It is equally important to understand what these arguments are not: allegations of bad faith. Most scientific advancements come about through this sort of give and take and the supplanting of one method with a superior method. My dissertation supervisor, Philip M. Hauser, who spent the early part of his professional career in the Census Bureau, used to say to me that the greatest compliment for a professor was to be proved wrong by one of his own students.

Most importantly, the fact that scientists argue among themselves over the proper way to count a rural area, or to inject noise into a database, or to run a particular analysis is not a sign that any of them seeks to hoodwink the public. Nor is it a sign that they are incompetent; these tasks are genuinely difficult and the "correct" course of action is not always evident. The arguments are not indicators of a hoax or misconduct. Impugning motives based on an honestly held professional opinion is a mistake typically made by politicians who do not understand the issue, or who seek to turn the issue to political gain for themselves.

The Census Bureau does a massive job of checking data quality and reporting on it. In effect, they are reporting on mistakes. This is not a sign that the Census Bureau professionals are somehow incompetent or not up to the job. Part of their job is to seek improvement, and the best way to seek improvement is to understand where a tabulation or analysis has fallen short of the mark.

Procedures in every census change in comparison with the preceding census. This happens because the standards for quality evolve. A larger, more diverse, and geographically dispersed population presents continuing challenges for those who undertake the decennial account. The Census Bureau works hard to document and explain the changes and then to measure the effects that the changes have.

---

[3] The apportionment formula itself aroused great controversy in the 1920s (Anderson 2015, pp. 150–53). Anderson proposes that the relative political calm of the 1950s, for example, brought attention to the issue of malapportionment. Anderson writes that reformers "pointed to two cogent reasons for reversing the trend toward malapportionment. First, they argued that by the 1950s, malapportionment was not simply a rural versus urban issue. The exploding suburban populations of middle-class whites were also underrepresented in Congress and their state legislatures, and the discrepancies were widening. Further, commentators in the 1960s suggested that the problems of the cities were at least partly the result of the decades of systematic underrepresentation and underfunding of urban areas" (Anderson 2015, p. 214).

The transparency of the Census Bureau in publishing evaluations of quality and estimates of error is itself a refutation of any notion that the census statistics are "fake news." The notion that a cynical bureaucrat is sitting in some back room plotting how to fudge an analysis is an unworthy caricature. What is more likely is that the statistics used in quality evaluations are unfamiliar to most people and hard to read for those not familiar with them. The Census Bureau tries diligently to explain technical materials in ways that the public can understand, but sometimes being accurate in the description of results also manages to obfuscate the results for some readers.

Madison saw constitutional checks and balances as the answer to unfettered partisanship. Ultimately the federal statistical system is not protected by any constitutional checks and balances. This is because a single political party in power in both Congress and the Presidency could order census funding, operations, or analysis in ways that further partisan ends. A simple example of this would be an amendment to Title 13 to permit sharing census information with law enforcement—a change that today could not be enacted but is an example of how the conditions of census taking could be radically changed by politics.

There are to be sure the checks and balances between the two branches. Congress is charged by the Constitution with the census, and Congress still has oversight over the process. In practice, however, the Executive Branch has been delegated responsibility for ensuring that the census is taken. The appropriation must come from Congress, and the Secretary of Commerce is named by the President and reports to the President. These provisions do not ensure that the census is free from political manipulation.

Ultimately the Constitution envisions the branches of government—Congress and the Executive—acting as checks on each other. This natural jealousy of the powers of one's own branch could be subordinated to party discipline. A possible scenario is that if a single political party controlled both of these branches of government, the party could pursue an agenda to subvert the statistical program. Perhaps the most subversive way to do this would be to use confidentiality as an excuse to avoid publishing the census or publishing some of the more interesting findings. This effort would disable the public from knowing or questioning the census results. While it would never be labeled as such, this would be a program of anti-transparency.

Today transparency is an important safeguard of statistical integrity. The provision of micro-data files and the extensive public tabulations allow the public to make their own interesting findings about the census, to detect errors, and to make their own contribution to improving the census.[4]

Ultimately, however, the guarantee of integrity in the census will not be constitutional checks and balances; as I have explained, there are circumstances in which they might not be sufficient. Nor is it the interaction with the external com-

---

[4] I have a very minor example of this process of correction in my work. In my dissertation research in 1973, I discovered a minor error in the magnetic tapes for one of the 1970 PUMS files. The Census Bureau promptly corrected it.

munity and the public; as I have noted, hypothetically that interaction could be curtailed or eliminated.

No, ultimately the guarantee of integrity in the census is the personal integrity of a host of talented, unsung, diligent professionals who work daily, whether it is a year ending in zero or not, to ensure that the U.S. Census is the very best it can be. Their work maintains an important pillar of U.S. democracy.

Please complete your 2020 Census form.

Glossary

Administrative records approach A technique by which administrative records are used to provide information from missed households or for missing data on incomplete household records. The viability of this practice is being evaluated by the Census Bureau.

Age heaping The tendency of people to round off their ages to numbers ending in 5 or 0 rather than providing exact ages.

American Community Survey (ACS) An annual survey mailed to over 3.5 million households by the U.S. Census Bureau that asks for detailed information about U.S. housing and population.

Bottom-coding Data points whose values below a lower threshold are censored to ensure confidentiality. See also, top-coding.

Canvassing The checking/counting of every address on every block in advance of the census.

Census The record resulting from the count of a country's population.

Census block The smallest geographic level in the census.

Census edited file A file released by the Census Bureau after the 2020 election that uses administrative records and statistical modeling to fill in missing data from the 2020 census. It contains sufficient geographic detail to be used in redistricting of a state.

Census unedited file The first file released by the Census Bureau after the 2020 Census. It will contain U.S. population counts by state. It is used for reapportionment of the House of Representatives among the states.

Characteristics imputation A statistical method for supplying missing data in a census return. An example would be if a household fails to indicate Hispanic origin for the members of the household. Normally there will be a telephone follow-up to get the information, but if follow-up attempts fail, then the information could be inserted into the file using administrative records or through one of several statistical practices. See, hot-deck editing.

Children The offspring of a person and/or the adopted children of a person. In most cases in the U.S. a child is anyone under 18 years of age, although there

© Springer Nature Switzerland AG 2020
T. A. Sullivan, *Census 2020*, https://doi.org/10.1007/978-3-030-40578-6

are some exceptions. Children, especially children below the age of 5, are often undercounted in the U.S. Census.

Citizen A person who has all of the rights and privileges a nation bestows to its people.

Citizen voting-age population (CVAP) data In response to President Donald Trump's Executive order 13880, CVAP will combine administrative data from a number of federal agencies into a separate microdata file that will contain a "best citizenship" variable for every person in the 2020 Census.

Citizenship question A question about the citizenship of a respondent. As an issue, it refers to efforts by President Donald Trump's administration to add a citizenship question to the 2020 census.

Community partnerships One of the Census Bureau's four strategies to increase the numbers of hard-to-count populations counted in the U.S. Census. Community partnerships include local elected officials, community leaders from non-government organizations such as churches and civic groups, and health and education leaders. It is especially useful to partner with members of HTC populations who can provide insight and even accompany enumerators, perhaps acting as translators or explaining cultural sensitivities.

Complete Count Committees Committees of elected officials, volunteers, officials from non-profit groups, and others in a position to help improve the response rate to the U.S. Census. Complete Count Committees may be organized at the state, county, or city level. Many states have provided appropriations to assist the work of Complete Count Committees.

Confidentiality A practice by which respondents' names and other identifying pieces of information are removed from their census returns. Data on the census returns are used only for statistical purposes. Confidentiality also implies a commitment to keep access to the data limited to individuals with a right to know.

Continuous Population Register As used in some European and Asian countries, the Continuous Population Register tracks every member of a population from birth until death.

Count imputation Imputing people for the population count. There are three types of count imputation. Status imputation is deciding whether a structure is also a housing unit. The second type of count imputation is occupancy imputation, which is deciding whether a housing unit is occupied. The third type of count imputation is household-size imputation, which estimates how many people live in an occupied housing unit when no household census return is available.

Coverage error An error in the census that results from not counting all members of the population, counting some members more than once, or counting members of the population erroneously.

Cracking A form of gerrymandering that fragments strengths of the opposing party by drawing a district line through their area of residential strength, then appending those areas to different districts that contain larger numbers of voters from one's own party.

Cyberthreat A risk or vulnerability that could be used to breach cybersecurity and cause harm or damage.

Data swapping A way that Census statisticians protect the identity of individuals or housing units with unique characteristics. For example, the unique characteristics of a household might be "swapped" with those of another household, while leaving other characteristics the same.

Deferred adjudication for childhood arrivals (DACA) DACA is a program created by President Barack Obama's executive order in 2012. The DACA applicants were foreign-born but had been brought to the United States as children, when they were too young to have any say concerning their entry into the United States. The parents who brought them were illegal entrants or became illegal later, typically through visa overstaying. DACA applicants have permission to remain and work in the United States under a set of strict provisions.

De facto counting The counting of a person wherever they actually are on census day.

De jure counting The counting of a person at their place of usual residence, even if the person is temporarily elsewhere on census day.

Demographic analysis A method used by demographers to ascertain whether an undercount has occurred in the census. Census data are compared with the count that would be expected based on other sources of data, such as birth and death records.

Design and accommodations One of the Census Bureau's four strategies to increase numbers of hard-to-count populations counted in the U.S. Census. These are minor variations from traditional census procedures meant to increase the response rate from hard-to-count populations. For example, the 2020 Census will be available in 13 different languages to help those who live in the United States but do not speak or read English.

Differential privacy A procedure that injects controlled noise into data in a manner that preserves the confidentiality of the respondent and preserves the accuracy of the data at higher levels of geography. This technique will be used in the 2020 Census.

Differential undercount The disparity between the net undercount rate for a particular demographic or geographic domain and the net undercount rate either for another domain or for the nation.

Editing (Along with Imputation) This involves the use of statistical methods to deal with missing information to yield a better count. Editing refers to changing answers to a census form that are internally inconsistent.

Electoral College A body of electors that represent each state in the presidential election. Their job is to formally cast votes for their state. Each state has the same number of electors as the sum of the state's Senators and Representatives. Thus, when census data are used to reapportion the House of Representatives, the Electoral College is also being reapportioned.

Family reunification Amendments made in 1965 to the Immigration and Nationality Act provided a preference for family reunification, so that many immigrants receive visas to come to the United States to join family members.

Freedom of Information Act (FOIA) A federal law that requires the full or partial release of documents controlled by the U.S. Government. Most states have a similar law applying to state records.

Gerrymandering Refers to drawing districts to further the interests of one party, and it carries the connotation of being unfair by giving one's own political party the advantage and disadvantage the opposing party. See also "packing" and "cracking."

Hard-to-count population A population that is more likely than others to be undercounted in the U.S. census because they might be hard-to-locate, hard-to-contact, hard-to-persuade-and/or hard-to-interview.

Hispanic origin People who identify themselves as Hispanic or Latino. The Census Bureau will offer the following categories for Hispanic Origin: Mexican, Mexican American, Chicano, Puerto Rican, Cuban, Latino or Spanish. People from other Latin American countries will be able to write in their origin. People of Hispanic origin may be of any race.

Homeless A person or family without regular or adequate housing, or who reside at a space not designed or used for sleeping, or who reside in a public or private temporary shelter, or who are exiting an institution where they temporarily resided.

Hot-deck editing Practice is based on the principle of homogeneity, or the fact that people living in a small area tend to be similar to one another. In this practice, the computer will be programmed to find the last complete household record from the same census tract that matches the incomplete household record. Then the incomplete record will be assigned the values that had been recorded in the complete record. Overall, the percentage of cases that receive this type of editing is small, and because it is small it is likely that it has little effect on the uses of the census data.

Illegal alien A person who is living in a country where they are not a citizen and does not have legal documentation to enter or to stay.

Immigration Immigration is the entry of a person born elsewhere into the United States for the purpose of permanent residence.

Imputation (Along with editing) A statistical method to deal with missing information to yield a better count. Imputation refers to completing incomplete questions or returns by reference to other information that might be available, such as administrative records. See also, count imputation and characteristics imputation.

Indirect disclosure Identification of an individual or household that occurs when legally available census data can be analyzed or linked to non-census data to reveal uniquely identifying information. Indirect disclosure is unintended on the part of the Census Bureau and the disclosure avoidance program is designed to prevent it.

Minorities People who are identifiable on the basis of race, origin, language, or other demographic characteristics and may suffer marginalization as a result. Marginalized people are often undercounted in the U.S. Census.

National Origins Act of 1924 This law that restricted immigration to the United States; it allocated entry based on the proportion of the U.S. population from a specific country in the 1890 Census. By using a census that was more than thirty years old, the Act had the effect of discriminating against the more recent immigrant stream from southern and eastern Europe.

Online response For the 2020 census, people living the United States will be able to complete their census form online.

Overcount Counting some members of a population more than once.

Packing A form of gerrymandering where all of the members of the opposing party are crowded into as few districts as possible.

Partial Population Register Administrative records that consists of data collected for a particular governmental purpose, such as conducting a particular program. These records are partial because they do not cover the entire population, but they can prove valuable for statistical purposes. For example, driver's license records are partial population registers. There are partial population registers at both the state and federal level.

Post-enumeration survey (PES) A sample survey taken shortly after the census. The respondents for the survey are not pre-selected; rather, they are selected through a multi-stage sampling process that approximates an equal chance for each household to be included.

Privacy Privacy refers to information that a person wishes to keep secret. Violating the sphere of privacy is perceived as intrusive, and many individuals find intrusive questioning by the government to be particularly offensive.

Processing errors Any inaccuracies that come from the ways in which the census data are collected, stored, coded, or analyzed.

Public-use microdata samples (PUMS) A sample of anonymized census or survey records available for analysis by the public. The Census Bureau produces PUMS files so that custom tables that are not available through summary data products can be created.

Racial categories The U.S. Census Bureau collects data about race through self-identification by census respondents. Respondents can self-identify as white, Black or African American, American Indian, Alaskan Native, Native Hawaiian, or Other Pacific Islander, along with Hispanic, Latino, or Spanish. Respondents can also answer Some Other Race or choose more than one race.

Reapportionment The redistribution of seats in the U.S. House of Representatives based on changes to population as recorded by the census. The U.S. House of Representatives has 435 members representing the 50 states, and each state's number of representatives is based on the state's population. Every state is guaranteed at least one Representative.

Record linkages In administrative and governmental records, identifying the records for the same person and providing means for the information in multiple files to be used.

Redistricting The process of drawing and redrawing electoral district boundaries in the United States.

Regression analysis A multivariate statistical procedure used to control simultaneously for a number of variables and estimate their relative effects.

Re-identification Combining information from different databases to identify someone in a census record. In recent years the Census Bureau has become concerned that the demographic and geographic data they release could be combined with other data to identify a specific individual.

Renters People who pay money to the owner of a housing unit to use the housing. Renters are often undercounted in the U.S. Census.

Respondent burden Also known as survey fatigue. Respondent burden refers to the time and effort required to complete requests for information, whether from the Census Bureau or elsewhere. Respondent burden is believed to increase non-response rates.

Respondent error Errors in self-reporting; the mistakes that people make, intentionally or not, in completing their census forms.

Response error See Respondent Error.

Rural residents People who live in rural areas, defined by the Census Bureau as places with fewer than 2500 inhabitants. Rural residents are likely to be undercounted in the U.S. Census.

Sample A mathematically chosen fraction of a population that is representative of the population from which it is drawn.

Sampling error An error in an analysis that results from unrepresentativeness of a sample.

Special programs One of the Census Bureau's four strategies to increase numbers of hard-to-count populations counted in the U.S. Census. These are major departures from the usual census-taking methods that are needed because of special needs of hard-to-count populations. For example, in the event of a natural disaster, a special program will be needed to count those persons living in the United States who might be displaced from their usual residence and residing in shelters.

Staff training One of the Census Bureau's four strategies to increase numbers of hard-to-count populations counted in the U.S. Census. This training is based on previous studies of the hard-to-count population, demographic analyses of the undercount, and analysis of changes since the previous census. The training provides enumerators with practice dealing with simulated difficult situations.

The Dillingham Commission A joint committee of U.S. Congress that concluded in 1911 that that northwestern European immigrants were more desirable than those from southern and eastern Europe.

The Monroe Doctrine A policy in the United States announced by President James Monroe in 1823 that sought to limit European colonialism in the Western hemisphere.

The United Nations (U.N.) Founded in 1945, the United Nations is an organization made up of representatives from 193 nations that is dedicated to a range of international issues.

Top-coding Data points whose points above an upper threshold are censored. See also, bottom-coding.

Undercount Counting less of a population than actually exists.

U.S. Bureau of the Census (Census Bureau) A non-partisan government agency responsible for amassing data about the United States and the people who live there. The Census Bureau is part of the U.S. Department of Commerce.

Wave election An election in which one party performs dramatically better than it previously had performed.

Index